# Tahoe Fracture Clinic's Guide to Joint Replacement

by
Martin Anderson M.D. & Jacob Anderson

Copyright © 2006 by Martin Anderson M.D.
All rights reserved. No part of this publication may be copied without the express consent of the author.

Anderson, Martin M.D., 1952 -
Anderson, Jacob, 1981 -

ISBN 978-1-4303-1673-2
1. Arthritis, 2. Orthopedic Surgery, 3. Total Joint Replacement

Published by Lulu Press
www.lulu.com

# Contents

Chapter I. Arthritis and Its Treatment — 1

Chapter II. Planning for a Joint Replacement — 31

Chapter III. Blood Use in Total Joint Replacement — 43

Chapter IV. Details of Total Knee Replacement — 51

Chapter V. Details of Total Hip Replacement — 79

Chapter VI. Medical Issues after Joint Replacement — 109

Chapter VII. FAQs (Frequently Asked Questions) about Joint Replacement — 119

Chapter VIII. Revision Surgery — 131

Chapter IX. Fractures around Total Joint Implants — 137

Chapter X. Total Joint Infection Surgery — 143

Chapter XI. A Primer on the Materials Used in Joint Replacement — 151

Appendix — 159
Glossary — 175
References — 185
Trademarks — 201
Index — 203

# *List of Tables*

| | |
|---|---|
| Table 1 - Common Types of Arthritis and Associated Diseases | 5 |
| Table 2. Grade Card for Osteoarthritis Medications | 17 |
| Table 3. Anesthesia Questionnaire | 38 |
| Table 4. Major Risks of Knee Replacement | 55 |
| Table 5. Major Risks of Hip Replacement | 81 |
| Table 6. Risks of Revision Surgery | 136 |
| Table 7. Anti-Inflammatory Medications Used for the Treatment of Osteoarthritis | 159 |
| Table 8. When to Stop Immunosuppressive Medications Before Surgery | 160 |
| Table 9. Comparative Risks of Blood Transfusion | 161 |
| Table 10. Health Information | 162 |
| Table 11. Knee Society Score | 165 |
| Table 12. Modified Harris Hip Score | 168 |
| Table 13. Tests to Evaluate Excessive Clotting | 169 |
| Table 14. Lab Abnormalities in Total Joint Patients with DVT/PE in Spite of DVT Prophylaxis | 169 |
| Table 15. Quantified Risks for Infection in Selected Conditions | 170 |
| Table 16. DVT Prophylaxis in Total Hip Replacement | 171 |
| Table 17. DVT Prophylaxis in Total Knee Replacement | 171 |
| Table 18. Consumer Reports of Glucosamine and Chondroitin Amounts | 173 |

# *Chapter I. Arthritis and Its Treatment*

## What is arthritis?

Arthritis literally means "joint inflammation." For types of arthritis not typically associated with significant inflammation, such as osteoarthritis, we are beginning to use the term "arthropathy," which means "joint disease." This is a guide to patients with arthritis of the hips and knees, particularly osteoarthritis (osteoarthropathy).

There are many acceptable ways to cope with, or even mitigate the symptoms of arthritis, but osteoarthritis has no cure. It can progress slowly or not at all, and it can be caused or worsened by injuries, including prior surgeries. Archeological studies show that arthritis has always affected human beings, and evidence indicates that it affects some races and some families more than others.

Arthritis affects 45 million people in the United States. There are over 150 known types of arthritis, which together affect one third of all adults to some degree. The most common form, osteoarthritis, is a disease whose cause is unknown. Arthritis affects the elderly more commonly than the young yet some elderly people do not have arthritis and some young people have severe arthritis. Osteoarthritis is a disease of the cartilage and bones that leads to pain in the joints. It is a degenerative process that is highly associated with obesity [7] but not with distance running [146]. Osteoarthritis does not seem to be simply a matter of wear and tear.

Many patients confuse the terms osteoarthritis and osteoporosis. Osteoporosis is a painless disease which leads to bone wasting. This disease—caused by hormone changes in middle age, inactivity, certain prescription drugs, smoking, and excess alcohol consumption—makes bones brittle and leads to fractures (broken bones). Osteoporosis doesn't cause pain unless a bone breaks. Osteoporosis does not prevent good results with joint replacement surgery.

Some types of arthritis are related to autoimmune disease. Rheumatoid arthritis, for instance, is an autoimmune disease which causes the cells that line the joints to attack the cartilage cells that provide painless joint motion. The treatment of autoimmune diseases may involve the use of medicines that alter the immune system and can have serious side effects, including infections. Joint replacement surgery should be postponed for two years after any infection in the arthritic joint. Due to the complexity of these diseases, there is a subspecialty of internal medicine, called rheumatology, which is dedicated to their treatment.

Osteoarthritis is a disease of the cartilage, but it eventually causes changes in the bone such as bone loss, bone spurs, and cyst formation. There is no cure for osteoarthritis and there is no way to accurately predict how fast it will progress. Only a small percentage of patients will develop arthritis severe enough to require surgical intervention. When osteoarthritis is this severe it is not easily confused with other diseases.

The approach that orthopedic surgeons take to arthritis depends on the severity of the disease. Physical therapy, activity modification, and the use of safe over-the-counter medications as needed (such as glucosamine preparations and acetaminophen), are often recommended for mild symptoms. This may mean giving up some activities, at least temporarily, while the cartilage that lines the joint recovers, and then gradually reintroducing those activities when the symptoms subside. Early arthritis responds to gradual strengthening of the muscles around the affected joint. Exercise with a physical therapist's supervision can help ensure that the chosen exercises don't cause further damage to the arthritic joint. Simply getting a gym membership and working with an athletic trainer may not suffice. Aggressive resistance training and high impact exercise may result in further damage to the diseased cartilage.

When symptoms worsen, but arthritis is still in its early stages, your orthopedic surgeon will examine the affected joint for evidence of physical derangements such as injuries to the cartilage, and for other mechanical derangements such as malalignment of the joint or loss of blood supply to the bones. In the knee, for instance, degenerative meniscus (buffer cartilage) tears may cause locking and collections of fluid (effusions), and these symptoms can usually be effectively treated with outpatient arthroscopic surgery. Conditions that predict a good outcome with arthroscopic surgery include mild to moderate arthritis (not bone on bone x-rays), locking, recent injury, swelling, and no prior knee surgery. Although arthroscopic surgery is safe, it is recommended neither for mild symptoms, nor in the setting of advanced arthritis. For symptoms that are not caused by a mechanical problem, short term use of anti-inflammatory medicines may be recommended. However, even these medications are associated with risks, particularly if they are combined, used for prolonged periods of time, or taken in excess.

For moderately severe arthritis, arthroscopic surgery is more controversial. Injections and long-term medication use may be considered. At this stage of the game arthritis sufferers must consider some lifestyle changes, such as aggressive weight loss strategies and completely giving up some activities to cope with their arthritis. Patients should substitute activities that result in pain and swelling with activities that do not result in these symptoms. As aerobic exercise is essential for cardiovascular and bone health, patients should find alternate activities that they can tolerate. Many patients will have little difficulty with swimming, walking, golfing, hiking, and bicycling even though they had to give up running, skiing and court sports. If arthritis symptoms become an excuse to become sedentary, invasive interventions such as surgery are preferable to becoming deconditioned and obese.

For severe arthritis with symptoms, options become more limited. Major surgery is often the only way to maintain activity at an acceptable level. As joint replacement (arthroplasty) surgery is the most successful reconstructive surgery in this instance, this book focuses on the needs of this group of patients. Orthopedic surgeons are now replacing 245,000 knees and 138,000 hips in the United States each year. These numbers are expected to increase to 454,000 knees and 248,000 hips by the year 2030. Mortality rates in the 90 days after total joint arthroplasty are about 0.7% in knees and 1.0% in hips in the population of Medicare patients[1,2], predicting a loss of life in 3095 of the 383,000 patients undergoing first-time joint replacement surgery this year. Based on annual death rates, 90-day death rates for people over 65 range from 0.45% yearly (65-69 year olds) to 3.7% (in those over 85 years old) [3]. Each year 16,500 people die from bleeding ulcers from the use of anti-inflammatory medications used to treat arthritis [4]. Joint replacement surgery is more effective than medical managment of severe arthritis and is relatively safe.

This pamphlet is written from the perspective of an orthopedic surgeon who specializes in joint replacement with the intention of helping patients understand the nature and risks of joint replacement surgery as well as the unique technical and economic issues that surgeons face when performing this surgery.

**Table 1 - Common Types of Arthritis and Associated Diseases**

| Disease | Comments |
|---|---|
| Osteoarthritis | Normal immune system. Causes knobby knuckles and loss of cartilage with bone spurs in multiple joints. Very common. |
| Post-traumatic arthritis | Similar to osteoarthritis but involving only an injured joint. May relate to an enzyme (metalloproteinase) which is produced for months after an injury or surgery. |
| Rheumatoid arthritis | Autoimmune disease where the cells lining the joint (synoviocytes) destroy cartilage. Symptoms include prolonged morning stiffness, nodules in fat, continuously swollen joints, and instability of neck vertebra. Leads to deformity of hands and feet. Treated with high doses of anti-inflammatory drugs and medicines that alter immune system responses. |
| Enteropathic arthropathy | Associated with Crohn's disease and ulcerative colitis. Treated with antibiotics, anti-inflammatory medications, and drugs that alter immune system responses. |
| Systemic lupus arthritis | Diffuse immune abnormality which can cause neurological disease (depression, psychosis, nerve injury, stroke), kidney disease (glomerulonephritis), chest pain with breathing (pleurisy), a rash over the nose and cheeks, and broken hair with balding. Treated with drugs that alter the immune system response. |

**Table 1 - Continued**

| | |
|---|---|
| Reiter's syndrome | Associated with venereal disease (chlamydia) or infectious diarrhea. May cause rash on palms of hands, soles of feet, and on penis (and is therefore often missed in women). Leads to arthritis of the hips and the sacroiliac joints. Treated with antibiotics and anti-inflammatory drugs. |
| Psoriatic arthropathy | Associated with psoriasis, a skin and nail condition. |
| Lyme disease | Associated with infections from tick bite and may lead to neurological problems such as Bell's palsy of the facial nerves. Treated with antibiotics and anti-inflammatory drugs. |
| Avascular necrosis (AVN) | Loss of blood supply to the bone that can eventually lead to bone collapse and severe arthritis. Associated with excessive alcohol intake and steroid use (such as prednisone). AVN may be treated with bone decompression or grafting or with medications called bisphosphonates that strengthen the bone to prevent collapse [5]. Small AVN lesions may resolve spontaneously while larger lesions often progress to bone collapse and severe arthritis even with surgical treatment. This disease is also called osteonecrosis. |
| Septic arthritis | Joint destruction from bacterial infection. Treated with drainage surgery and antibiotics. Joint replacement surgery should be postponed for two years after infection. |

# Treatments for osteoarthritis

## 1. Activity modification

Patients suffering from osteoarthritis should limit their activities prudently, but should not give up exercise as it is important in maintaining cardiovascular fitness and bone health as well as emotional and physical well-being. It may be prudent to substitute walking or bicycling for jogging if you are developing arthritis. Swimming and water aerobics are better tolerated for patients with severe arthritis of the weight bearing joints. In all instances exercise should be prudent, with adequate periods of rest. Listen to your body: if an exercise consistently results in pain and swelling, substitute another activity in its place. There is a body of evidence that suggests that weakness of the muscles around the arthritic joint may accelerate the progression of arthritis. Therefore, rest should only be used to treat symptoms for short periods of time, followed by reintroduction of activity to prevent weakness from compounding your problems. The bottom line is that it is the role of physicians and surgeons who specialize in the treatment of arthritis, to keep you active.

Cardiovascular conditioning should be discussed with your primary care provider, especially if you have a history of heart problems or risk factors for cardiovascular disease such as smoking, high blood pressure, diabetes, high cholesterol, obesity, or a family history of heart disease under the age of sixty years old.

## 2. Physical Therapy

Physical therapy typically consists of strengthening, flexibility work, aerobic training, and physical modalities such as electrical stimulation, ultrasound, and acupressure. While there needs to be additional scientific work done to clarify the success of physical therapy in managing arthritis, a panel of British experts recently evaluated the evidence-based literature on the role of exercise in a study called the MOVE consensus [6]. There is scientific evidence that strengthening and aerobic exercise can reduce pain and improve function in patients with knee arthritis. The evidence that therapy helps hip arthritis patients is lacking but experts believe that it will help, extrapolating data from studies of knee arthritis. Worse outcomes can be anticipated with physical therapy for patients with the most severe arthritis, as gauged from x-rays. Strong evidence exists that home exercise programs can be just as effective as group exercise programs and that a key role of the therapist is to provide advice and education.

## 3. Weight loss

Obesity is a major cause of arthritis [7]. The additional weight places greater stresses on joints and the sedentary lifestyle that may have led to obesity has been associated with more rapid arthritis progression. Weight loss will significantly benefit overweight patients as many activities of daily living place several times the body weight across portions of the knee and hip. Although the benefits of weight loss are exponential, there has not historically been a safe and effective medical treatment for obesity. Diet and exercise are more successfult than either alone and benefit people with arthritis[7]. There are new surgical treatments available for patients who are morbidly obese, although previous bypass and stapling operations were unsuccessful. Morbid obesity (over 100 pounds overweight) means that you are so overweight that your obesity is expected to hasten your death. The subspecialty of general surgery that provides surgical treatment of morbid obesity is called bariatric surgery.

Obesity, in itself, does not preclude a good result from arthritis surgery, but it is associated with higher rates of blood clots, joint instability, infections, and wound healing problems. The white blood cells that fight infection do not work correctly in obese patients. Also, the surgery is more technically demanding for your surgeon. Historically, many surgeons have refused to operate on morbidly obese patients and, even now, you should understand that many surgeons do not have the skill or stamina to operate on very large patients and may recommend that you seek surgical care from a sub-specialist. You may not lose weight after a total joint replacement [8]. Just because you are able to resume a more active lifestyle after surgery is no guarantee that you will become more active. It is hard to change the behaviors that you have adopted to cope with your arthritis, and if you have always been sedentary, you will need motivation other than a new joint to make you more active.

### 4.  Glucosamine and artificial joint fluid:

Glucosamine and Chondroitin Sulfate are over-the-counter derivatives of shells from shellfish that are sold as nutritional supplements. They are the only supplements that have been shown to slow down the progression of arthritis. There is experimental evidence that these medications improve the health of cartilage cells in animals but no evidence that they restore lost cartilage cells or make those cells reproduce faster. A study showed that long-term use of glucosamine did slow the progression of arthritis of the knee based on measurements of remaining cartilage made from x-rays and compared to use of a placebo [9]. Glucosamine has been shown to be as effective as over-the-counter ibuprofen in treating arthritis pain, and it is safer than anti-inflammatory medications. Only rarely are allergic reactions seen, and these typically occur in patients who are allergic to shellfish. Insulin resistance (a cause of diabetes) has been seen in animals given glucosamine and some diabetic patients have reported higher blood sugars when using glucosamine. The recommended dose is 1500mg daily.

The GAIT (glucosamine/chondroitin arthritis trial) study has released conflicting results [10]. While there is not yet statistical evidence that these compounds outperform a placebo, the group of patients with moderate to severe arthritis pain is benefited more by glucosamine and chondroitin sulfate than they are by Celebrex™. Longer follow up may further clarify the role of these nutritional supplements.

It is important to discuss brand choices with your pharmacist as these companies historically have not been required to guarantee that they have the correct amount of the supplement in their tablets, and some brands have no glucosamine in them [11]. Results for assays done be Consumer Reports in June of 2006 on various brands are listed in Table 18.

### Viscosupplementation (artificial joint fluid):

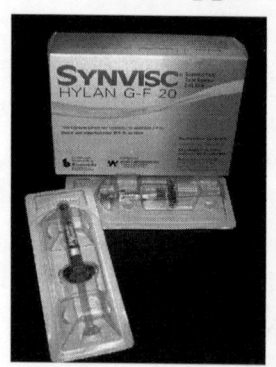

These prescription products are derived from rooster combs and used for injection into the arthritic knee. Early studies show that this artificial joint fluid improves pain in 70% of patients for an average of six months. However, these results are clouded by less optimistic results in other studies. The treatment, used primarily for the knee joint, consists of a series of three to five injections given roughly one week apart. While this is expensive and has not been shown to prevent arthritis progression, it may buy time before surgical options are considered. It may make remaining cartilage cells healthier but does not cure arthritis. The results from these injections appear to be better in patients who have not progressed to severe arthritis and who do not have serious mechanical problems, such as tears in the cartilage.

This treatment is expensive and should your insurance fail to approve the use of this medication your costs may be in excess of $1000. If your insurance won't cover this treatment or if your arthritis is severe (bone on bone), injection of cortisone-type medications may be more prudent: one study showed no difference in the results of artificial joint fluid injections compared to a less expensive cortisone injection [12].

Artificial joint fluid injections may cause a local allergic reaction. This occurs about 2% of the time with the first series of injections and 21% of the time with the second series with the use of Hylan GF-20 (Synvisc™) [13]. Allergic reactions may require removal of the fluid (aspiration) from the knee along with a cortisone-type injection. Allergies occur because these preparations are made from rooster combs and this material is foreign to our immune systems.

Infection risk is 1/10,000 with injections of the knee. An infection makes total knee replacement inadvisable for a period of two years due to the risk of recurrent infection. At the time of this writing, viscosupplementation treatment is FDA approved for the knee only. Its use in other joints is being studied.

## 5. Injection of "cortisone," numbing medicines, and joint aspiration

This treatment may actually decrease the disease activity of rheumatoid arthritis but it does not do so for osteoarthritis. Cortisone was first used to treat rheumatoid arthritis at the Mayo Clinic in the 1948 by Dr. Philip Hench, who shared a Nobel prize for reasearch on cortisone in 1950. It provides pain relief for hours to months, depending on how much inflammation is associated with the arthritis, and, like other medical interventions, it works well when arthritis is not severe (bone on bone). Although we still refer to these injections as cortisone, we don't actually inject cortisone anymore. We now use medications which are less soluble (don't dissolve as well). These drugs are of the same class of medications as cortisone (glucocorticoids), but they last longer. Patients often find that the first injection for osteoarthritis may last a few months. Subsequent injections don't last as long. Frequently the second injection lasts several weeks, the third lasts several days, and the forth injection lasts only several hours. Most surgeons do not recommend this treatment more than four times yearly, and no more than once a month. It carries an infection risk of about 1/10,000, and it may temporarily cause significant increases in blood sugar in diabetic patients. Cortisone injection for mild arthritis may not be advised due to our uncertainty about how the medications may affect the remaining cartilage. The hip joint usually requires injection by a radiologist using a fluoroscope. Injection of the bursa outside of the hip joint is a simple office procedure.

When there are questions about the source of pain, injection of the joint in question with numbing medicines can be valuable in helping your physician in confirming a diagnosis. This is called a diagnostic injection.

Aspiration of a joint is done to remove the fluid from the joint, also usually for diagnostic purposes. This can be useful to diagnose an infection or gout. Repeated aspirations of a joint are not recommended because the joint fluid rapidly returns when the joint is inflamed.

## 6. Acetaminophen (Tylenol™)

At doses not exceeding 4 grams daily, acetaminophen is a relatively safe treatment for pain. That is, you should avoid more than eight extra strength (500mg) tablets or twelve regular strength (325mg) tablets throughout the day. If you do exceed this dose, liver damage may occur, especially in heavy drinkers. Remember that many narcotic preparations (Darvocet™, Vicodin™. Lortab™, Percocet™ and others) contain acetaminophen and should not be supplemented with additional acetaminophen.

Acetaminophen was shown to nearly double the risk of high blood pressure (hypertension) in one study [14]. This study is part of the Nurses' Health Study which only includes women. This study demonstrated that headaches didn't cause hypertension unless acetaminophen was used to treat them.

Presumably, it was the acetaminophen and not the pain that caused the blood pressure problems. There is no reason to believe that the association between acetaminophen and blood pressure elevations will be limited to women although this discovery was made in this women's study.

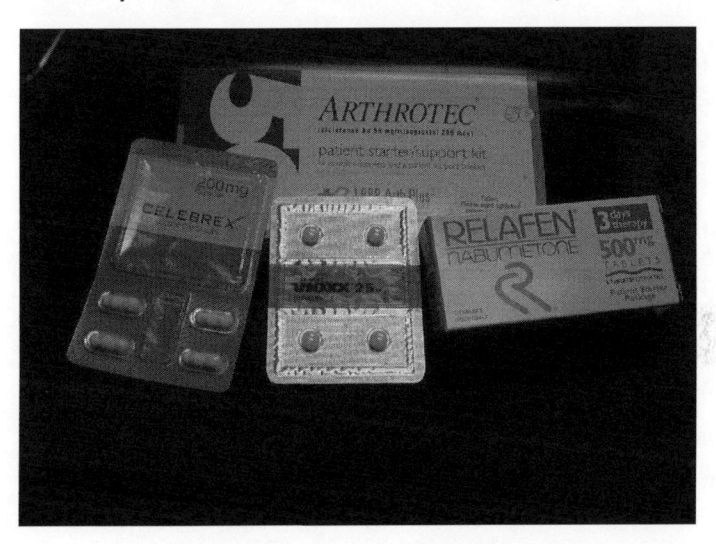

## 7. Anti-inflammatory medicines

Anti-inflammatory drugs are useful to decrease inflammation, shorten the duration of symptoms, and treat pain. Although these drugs are useful when inflammation does not play a constructive role, they may hinder healing when inflammation does play a constructive role, such as in bone fracture healing, after bone fusion surgery and, in bone ingrowth after cementless total joint replacement.

These medications may decrease pain and inflammation when it occurs in osteoarthritis, but this class of medications carries the risks of potentially fatal bleeding ulcers, high blood pressure, swelling, liver damage, heart attacks, strokes and kidney failure. The use of these medications will require laboratory monitoring. Patients with a history of ulcers or aspirin allergies should avoid this class of medicines.

A reasonable first step with anti-inflammatory medicines is to try over-the-counter medicines, such as ibuprofen or naproxen. You should discuss their use with your physician who can determine if they are safe given individual circumstances. If these are ineffective or if longer term use is anticipated, prescription medications such as combinations of anti-inflammatory medications with medications used to treat or prevent ulcers may be considered as the next step to decrease the risk of bleeding ulcers.

Each of the anti-inflammatory medications work to inhibit an enzyme in the body called cyclooxygenase (COX). There are two sites. COX1 receptors are found in the stomach, kidneys, liver, and the platelets (which initiate blood clotting). COX-2 receptors are found in the bones, joints, tendons and also in the kidneys. The development of COX-2 selective and COX-2 specific drugs allow physicians to target the enzyme in musculoskeletal tissues better and decrease ulcer rates compared to non-selective medications. Selective COX-2 inhibitors wouldn't be expected to protect the kidneys, however. The medicines Mobic™, Relafen™, and Lodine™ are COX-2-selective and may be safer on the GI tract than older medications. They are relatively safe and effective. COX-2 specific drugs target the enzyme (cyclooxygenase) in the muscle even more precisely than COX-2 selective drugs.

Celebrex™ and Vioxx™ are COX-2 specific drugs. COX-2 specific medications reduce the risk of bleeding ulcers, which is the major source of death from anti-inflammatory medications. However, they carry a risk of coronary artery occlusion (heart attack) and stroke. Vioxx™ was voluntarily taken off of the market in 2004 because of these concerns. Non-selective anti-inflammatory medications also carry this risk to some extent.

**GI side effects of NSAID's (Nonsteroidal Anti-Inflamatory Drugs):** The older, non-selective medicines cause ulcerations in the GI tracts (esophagus, stomach, duodenum, or colon) in about 15% of long-term users and bleeding ulcers requiring hospital admission in approximately 1% of regular users annually. If they bother your stomach or cause abdominal pain, stop using them. If you have blood in the stool, black tarry stools, vomiting of material that looks like coffee grounds, or just feel ill, stop the medicines and seek medical care immediately. Remember that life threatening GI bleeding may occur without pain or other warning symptoms. Close monitoring for GI ulceration is required as it is estimated that 16,500 arthritis patients die from bleeding ulcers caused by taking these medicines in the U.S. annually [4]. The use of medications that heal ulcers may be recommended if you are prescribed anti-inflammatory medications as these medications may prevent ulcers.

**NSAID's can cause liver and kidney injury, high blood pressure and swelling.** Each of these medications can cause edema (swelling) and worsening of hypertension (high blood pressure) in some patients. Aspirin is the only medication from this class that is not believed to cause hypertension. Kidney failure or liver damage may occur although this is uncommon and usually occurs in patients who already have some degree of kidney failure, liver disease, or heart failure. If you have some degree of kidney failure, your doctor will need to order a blood test to calculate the risks of using these medicines.

**Bleeding risks:** The older medicines (those that aren't COX-2 specific) may interfere with platelets (blood clotting) and are unsafe before surgery and when the blood thinner Coumadin™ (Warfarin) is being used.

**Allergic reactions from NSAID's:** Patients with sulfonamide ("sulfa") allergies may have severe allergic reactions to Celebrex™ or Bextra™. Those with NSAID or aspirin allergies should not use any of these medicines if it is determined to be a true allergy. Those with asthma, urticaria (hives from environmental allergies), and nasal polyps should discuss the risks of these medicines with their doctor before using them. Stevens-Johnson syndrome, a rare and potentially fatal skin and mucous membrane disease, has been seen more frequently in patients taking Bextra™ than in those taking other NSAID's. Bextra™ was removed from the market due to this concern.

**Heart attack and stroke from NSAID's:** COX-2 selective inhibitors do not protect you from heart attack, stroke, or blood clots. This is expected as they do not thin the blood (they don't affect blood platelets). They change clotting mechanisms in complex ways, such as through the cells that line blood vessels, called endothelial cells. This is called endothelial dysfunction. If you have risk factors your doctor may advise you to take a baby aspirin to decrease these risks. Common risk factors for heart attack (MI) include: previous MI or stroke, known coronary artery disease, carotid or peripheral vascular disease, previous bypass or stent surgery, smoking, diabetes, obesity, worrisome family history, high cholesterol, and leading a sedentary lifestyle. The use of aspirin with COX-2 drugs negates much of the protective effect on the GI tract.

COX-2 inhibitors have been shown to introduce a higher risk of strokes and heart attacks. The use of nearly all anti-inflamatory medication, particularly with higher doses or prolonged use, share this risk to a lesser extent. Aspirin is the only anti-inflammatory medication without this risk, but aspirin can cause bleeding ulcers.

For long-term use of anti-inflammatory medicines physicians recommend periodic lab testing of their patients to ensure that no problems are occurring. Lab testing includes three stool hemetest cards which detect bleeding from the GI tract, kidney function tests and liver function tests. In most circumstances, these tests would be performed three months after starting the medicine and then at least once yearly. Your blood pressure should also be monitored closely as should the test for blood thinning (INR) if the blood thinner warfarin (Coumadin™) is used. If you are on warfarin, you need to

## Table 2. Grade Card for Osteoarthritis Medications (Authors' Opinion)

| Drug | Ulcer | Heart attack or stroke | Blood pressure increase | Swelling | Effectivness | Cost |
|---|---|---|---|---|---|---|
| Glucosamine | B | B | B | B | C | $$-$$$ |
| Tylenol™ | B | B | D | B | B | $ |
| Aspirin | F | A | B | B | B | $ |
| Non-selective NSAID with ulcer medicine | C | D | D | D | B | $$$-$$$$$ |
| Non-selective NSAID alone | F | D | D | D | B | $$-$$$ |
| COX-2 specific NSAID | D | F | D | D | B | $$$$ |

**Risk Grading System**
A=protects against disease
B=doesn't cause disease
C=mixed data
D=causes disease to a lesser extent
F=highly associated with disease

**Effectiveness Grading System**
A=curative
B=improves symptoms
C=mixed data
D=no better than placebo
F=worsens symptoms

have a blood test (INR) 3-5 days after beginning the use of anti-inflammatory medication. All of these medications are potentially unsafe when taken with warfarin.

The FDA recommends using the smallest doses possible for the shortest times possible and avoiding these medications entirely after cardiac bypass procedures. Most physicians would offer this same advice.

### 8. Topical preparations

Anti-inflammatory medications can be "complexed" into gels by the pharmacist, allowing some medication to be absorbed across the skin without the risk of systemic problems (except allergic reactions). For instance, the pharmacist can mix a 10% ketoprofen gel. These formulations have a short shelf life and won't work after they are old. There is training required for the pharmacists in this instance and you may not be able to use your usual family pharmacist for this prescription.

The medication, capsaicin, can both block the transmission of pain through nerve fibers and can block formation of a pain transmitter (substance P), thereby alleviating pain. Its use, however, is limited by its tendency to cause a burning sensation in warm weather or when exercising. Patches which numb the skin have also been used.

### 9. Pain medications

The prescription medication tramadol (Ultram™ and Ultracet™) is not believed to be as habit-forming as narcotics but there have been cases of addiction. These drugs can also be expensive, but tramadol is also available as a generic drug now.

Narcotic pain medications must be used with caution and their long-term use is often supervised by a physician who specializes in the treatment of chronic pain. Narcotics (such as morphine, Demerol™, Dilaudid™, Oxycontin™, Percocet™, Tylox™, Vicodin™, Lortab™, and Darvocet™) can cause addiction, constipation, depression, and mental status changes, and they typically require increasingly higher doses as your body becomes tolerant to them. These medications impair judgment and reaction time and must not be used within at least 8 hours of driving or operating machinery, depending on the medication. It is just as illegal to drive while taking these medications as it is to drive drunk. The decision about driving with chronic (long-term) use of narcotics must be made by your pain specialist, often a physiatrist (physical medicine and rehabilitation specialist). The use of a pain pill "now and then" is no safer than smoking a cigarette now and then from the standpoint of addiction. It can be a starting point for addiction. People with chronic severe pain are not believed to be as prone to addiction as others but arthritis pain is rarely severe enough to warrant narcotic therapy and addiction remains a concern in this group of patients.

Risks include: addiction, depression, nausea, vomiting, impaired judgement, slowed reaction time, confusion, hallucinations, allergic reactions, constipation, pseudo-obstruction of the bowel, bowel perforation, respiratory depression, and potentially fatal heart rhythms (especially with Darvocet®-type medications which have a toxic metabolite, nor-propoxyphene). Some of these risks can lead to death. Do not mix these medications with alcohol, sedatives, tranquilizers, or muscle relaxants unless your physician informs you that the drug combinations are safe.

## 10. Braces

Some people get relief from a wrap worn around the knee although there is no evidence that knee sleeves, with or without magnets, will improve their arthritis. So long as they are not worn so tightly as to cause swelling or blood clot formation, they are safe. Unloading braces are believed to benefit arthritis pain when the arthritis is limited to only part of the knee. Many patients ask their surgeons to prescribe these braces but too often the result is disappointing. Some scientific studies of these braces have shown improvement in the gait pattern but other studies have shown no improvement in pain relief compared to a simple knee wrap obtained at a much lower cost and without a prescription from the doctor. Custom-made arthritis braces are expensive, cumbersome to wear, and occasionally make the knee pain worse. However, some patients experience relief from these devices.

## 11. Non-traditional therapies

Acupuncture has been shown to benefit arthritis patients [15]. Alternative and complimentary medicine is a complex topic that is beyond the scope of the specialty of orthopedic surgery. Although the science supporting alternative and complimentary medicine research has traditionally been weaker than the science required for drug and medical device research, some patients appear to have success with these alternative therapies. Orthopedic surgeons tend to support whatever interventions that you and your nutritionist or naturopathic physician agree on, as long as they are safe. Most orthopedic surgeons have no personal expertise in this area and will not be able to make recommendations regarding complimentary and alternative therapy to you in good conscience.

## 12. Arthroscopic surgery of the knee

This works best for patients who have symptoms of short duration with mechanical derangements such as meniscus tears, central osteophytes (bone spurs), or loose bodies in the joint, affording at least two years of relief for 80% of patients that fit these criteria and do not have severe arthritis. Repeat procedures, arthroscopy for severe arthritis (bone on bone), and procedures for patients with symptoms which came on slowly, have persisted a long time, and who don't have meniscus tears is less rewarding. In a study performed by the Veteran's Administration [16], arthroscopic surgery for patients with severe arthritis was shown to be of no benefit. In this study, half of the patients had a sham operation where incisions were made but no arthroscopic surgery was done. The other half had the usual arthroscopic surgery. Neither the patients nor the doctor seeing them after surgery knew which group they were in (a double blind study). Surprisingly, there was no difference in the outcomes between these two groups of patients. The patients who received the sham (fake) operation had a result that was just as good as the patients who actually had the surgery. This study has caused us to reevaluate the role of arthroscopic surgery for knee arthritis to see if there are subsets of arthritis patients who still might benefit from this surgery.

Torn Meniscus

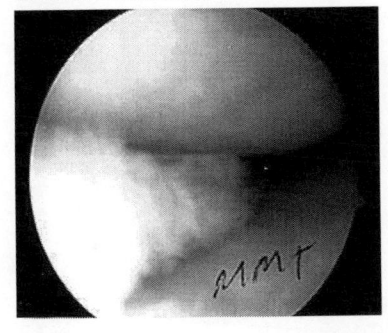

Tear Removed

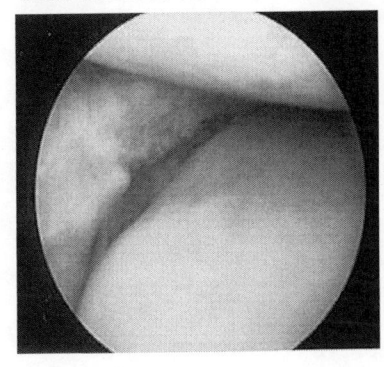

Surgery to replace cartilage (mosaicplasty and autologous chondrocyte implantation) is appropriate to repair holes in cartilage, not huge surfaces of absent cartilage such as occur in severe osteoarthritis. The problem is that cartilage cells do not adhere to bone as they would in a normal knee due to the loss of the supporting collagen framework present in normal joints.

Mosaicplasty (also called the OATS procedure) is a technique by which plugs of bone and cartilage are harvested from parts of the knee that are believed to be less important and pressed into similar sized drill holes inside an area of missing cartilage. Moderately large defects can be filled in this fashion by placing multiple plugs, forming a mosaic of new cartilage attached to underlying bone. To replace all of the bone and cartilage of the knee as would be required in severe osteoarthritis requires massive osteochondral allografts. "Osteo" means bone and "chondral" means cartilage. In this procedure, large portions of donor bone and cartilage are grafted onto the arthritic knee or ankle in a similar fashion to total joint replacement. Currently, there is not much data on these procedures.

Chondrocyte implantation involves implanting cartilage cells grown in a tissue culture into a hole in the cartilage. In the first stage of a two-stage procedure, an arthroscope is placed in the knee to remove cartilage. The cells are then grown in a tissue culture. In the second surgery, a patch of tissue from the thigh is sewn into place around the hole. The patch utilizes the cartilage at the periphery of the hole to anchor it. Cartilage cells are injected into the defect and are contained by the patch. This requires a rim of healthy cartilage around the defect to sew the patch into. Healthy cartilage rims are not commonly present in patients with severe osteoarthritis.

Another procedure, the microfracture technique, can be used to induce new scar cartilage (fibrocartilage) to form in small defects[17]. This procedure allows stem cells inside the bone marrow to differentiate into cartilage cells. It may work much better for younger patients as the population of stem cells which can make cartilage decreases with age, at least in women [18]. Microfracture technique has equivalent results with chondrocyte implantation [19]. This technique is not as successful for obese patients as it is for non-obese patients, however [20]. Microfracture technique requires much less surgery than chondrocyte implantation for repair of holes in the cartilage. Most surgeons prefer the microfracture technique to cartilage transplants because the microfracture technique can be done with simple arthroscopic surgical techniques and does not require removal of cartilage from other parts of the knee.

Consider the analogy of repairing a road: we can repair the potholes with cartilage ingrowth and transplantation technology, but we can't lay a whole new highway of cartilage. We can grow abundant cartilage in cell cultures, but without collagen to attach the cartilage to the bone this technology is helpless against severe arthritis. Arthroscopic surgery does offer many patients relief, especially if done before arthritis becomes severe.

Arthritis is not simply a collection of mechanical derangements that can be repaired with hand tools. There are changes that affect the cells that line the joint cavity (synovial cells) as well as the cartilage itself. Osteoarthritis is a breakdown of the control mechanisms of multiple types of cells that ultimately lead to the loss of cartilage substance. Research is now focused on the cellular and molecular biology of this process and in gene therapy.

To estimate the likelihood of a successful result with the arthroscopic surgery, the surgeon weighs the following factors:

### FACTORS WHICH PREDICT SUCCESS WITH ARTHROSCOPIC SURGERY [21]

Short duration of symptoms
Recent injury
No previous surgery to the knee
Locking of the knee
No pain at rest
No malalignment
Effusion (fluid on knee)
X-ray not "bone on bone"

It is less common to recommend arthroscopic surgery for arthritis in joints other than the knee and shoulder, but occasionally the ankle, elbow, wrist and the hip will benefit with arthroscopic surgery when arthritis is present. For a very young or very active patient with arthritis it may be worth the risks and limited goals of arthroscopic surgery to attempt to postpone joint replacement surgery even if there are factors that may predict limited success as outlined above.

## 13. Osteotomy (bone realignment) surgery

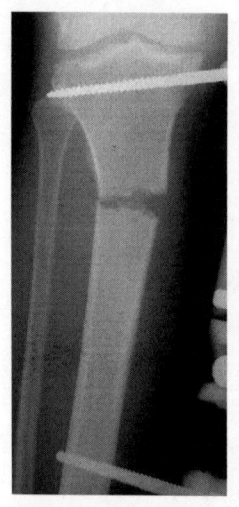

Osteotomy surgery is occasionally used for young patients with hip disease but is a much more common operation for adults with knee arthritis that is limited to one compartment of the knee. The goal is to realign the knee to allow more weight to be borne through the remaining healthy compartment. It may complicate future total knee replacement but remains an option for patients between the ages of 40-65 who are not obese, who have at least 90° of motion, and who do not have a severe deformity (over 15 degrees of bowlegged deformity, called varus deformity, for instance). It appears to have slightly better results in women and less active men, with 75% of all patients reporting a satisfactory result at 7 years [22]. It has the advantage over joint replacement of not requiring the implantation of an artificial prosthesis.

The major risks of this surgery are failure to relieve the pain in 25% of the patients, injury to the peroneal nerve with foot drop in 10% [23], infection in 1%, pin tract infection if an external fixator is used, failure of the bone to heal, scarring of the tendon below the kneecap resulting in a low-riding kneecap (patella infera), injury to the joint between the tibia and the fibula, and loss of the blood supply to the top of the tibial bone (avascular necrosis). The usual surgery results in a cosmetic deformity as it is designed to create a knocked knee deformity to allow more weight to be transferred to the less arthritic outer compartment of the knee. Other complications include injury to other nerves or blood vessels, blood clots in the legs which may travel to the lungs and become life-threatening, and failure of the hardware. Osteotomy surgery for knee arthritis is less common in recent times because of the high success rates and more durable results with total knee replacement.

Usually when the outer (lateral) compartment of the knee is selectively worn out, the end of femur (thigh-bone) is moved, and when the inner (medial) compartment of the knee is selectively worn out, the top of the shin bone (tibia) is moved.

Osteotomy of the bones around the hip joint can be very effective when the hip joint has failed to form correctly (congenital hip dysplasia) or when it has been damaged from certain childhood diseases. This surgery is done to halt the progression of arthritis and is usually not done after the age of 40 or in patients who have significant arthritis of the hip joint. Done well earlier in life, the osteotomy may make it easier to replace the hip later. There are few surgeons experienced at pelvic osteotomy surgery.

## 14. Fusion surgery

Hip fusion is a good operation for very young, very active patients with severe hip arthritis. When done correctly (with the hip in good position: flexed 30°, abducted 0°, and externally rotated 5°), it results in a minimal limp and provides relief for 20 years or longer. When it does fail, it usually does so by causing back and knee problems [24]. It may be possible to convert a fused hip to a total hip replacement years later. Hip fusion does limit the ability of a patient to spread their legs and this can be a detriment to sexual relations. Hip fusion may shorten the leg and require a built up shoe. Failure to fuse is not rare and often requires additional surgery.

Knee fusion surgery is also effective but most patients are annoyed by the fact that the knee is straight when sitting, requiring special seating arrangements on airlines and tripping other patrons in movie theaters. It usually results in enough shortening of the leg to require a shoe build-up. Knee fusion is associated with a more noticeable limp and is much more difficult to convert to a total joint replacement in later years than is hip fusion. Knee fusion surgery is still the recommended procedure for incurable bone infections especially when there is severe soft tissue loss.

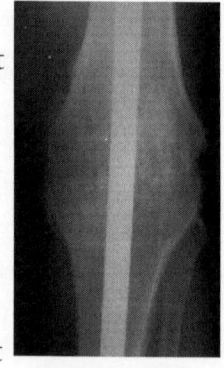

Fusion surgery is widely used to treat hand, wrist, foot, ankle, and spine arthritis. Fusion removes the motion from the joint but a successful fusion is almost always painless.

## 15.    Aids to ambulation

The hip joint is subjected to forces equal to nearly 3 times the body weight when walking, due to the biomechanics of its muscle actions. These forces can be offset as much as 40% by using a cane in the other hand [25]. A cane can benefit knee pain to a lesser degree but is handy to prevent falls should the knee buckle with pain. Use a cane in the hand opposite of your bad hip or knee.

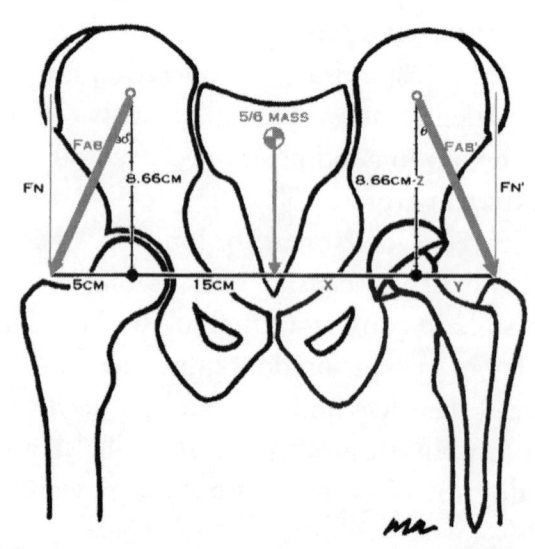

Because the knee or hip can give way (buckle) from

arthritis pain, falls can occur and may result in broken bones. Fractures greatly complicate the surgical treatment of arthritis. With risks for falls it is advisable to use a walker for additional support.

Wedged insoles may alleviate symptoms from knee arthritis. The wedge is typically placed on the outer part of the heel for arthritis of the inner part of the knee.

## 16. Total joint surgery

Total joint surgery is clearly the gold standard in the treatment of severe arthritis of the hip, knee, shoulder, and elbow. Results from replacement of the ankle, wrist, and toes are too variable in most surgeons' hands to recommend them in many instances, although special circumstances occasionally warrant their consideration. In general, more than 95% of all patients obtain good (minimal pain) or excellent (no pain) results. This still leaves a few total joint patients dissatisfied with the result to one extent or another, but it is uncommon to have a patient report that they are worse off than they were before surgery.

In a recent survey conducted by Consumer Reports, the results are slightly different than the results of joint replacement in surveys conducted by surgeons [26]. While 82% of patients were "very" or "completely" satisfied with the results from hip and knee replacement, 5% reported some type of infection and 13% reported a weakened muscle requiring more therapy. 13% of hip replacment patients and 7% of knee replacement patients did not feel that their leg lengths were equal after surgery. Complications were more common in obese patients and results were worse in patients who put their surgery off for long periods of time. Consumer Reports recommended finding a surgeon who specializes in hip and knee replacement and who does at least 50 of these surgeries a year.

A total joint replacement will eventually wear out and

require repeat (revision) surgery if you live long enough. Hip replacement surgery affords good pain relief early on. Knee replacement surgery will require several weeks to gain good pain relief. Pain and motion will continue to improve over the entire first year after knee replacement surgery.

Results from joint replacement vary from surgeon to surgeon and from hospital to hospital, but complications can occur in anyone's hands. The complication rates are lower among surgeons and centers that do more of this surgery than by those that only perform this surgery occasionally. The most common complications include: malalignment of the components which can lead to excessive wear or instability, changes in leg length, dislocations of the hip joint or kneecap, infections requiring at least temporary removal of the implants, potentially fatal blood clots, implant loosening, fractures of the bone, and continued pain or loss of motion with or without soft tissue calcification (heterotopic ossification). Please refer to the detailed sections that follow on hip and knee replacement for more information on these potential complications.

In most instances joint replacement is safe and cost effective, and nearly always makes it possible for patients with severe arthritis to lead more active lives. Joint replacement is not recommended for mild arthritis, when active infection is present, or when other active medical problems create unacceptable risks. Age is not an absolute contraindication to this surgery, which the author has done in patients as old as 102 and as young as 19. Patients over 80 years old are more prone to neurological, pulmonary (lung), and cardiac problems and often require additional testing before surgery and closer care after total joint surgery [27]. The surgery can be done reasonably safely even when there are other medical and cardiac problems if certain precautions are followed. Additional caution must be exercised in patients with cardiac, lung, liver and kidney disease as well as in patients with abnormal immune systems.

Joint replacement can often be performed in 45-90 minutes. The hospital stay is usually about 48 hours, but some patients will benefit from an additional stay in a rehabilitation hospital. This consideration varies depending on your insurance coverage and ever-changing CMS (Medicare) regulations. Called the "75% rule," there is increasing enforcement of a CMS (Medicare) rule that requires that no more than 25% of the admissions to a rehab hospital be patients who have had joint replacement surgery. The only total joint replacement patients that are exempt from this requirement are those who are over 85 years old, those who have both-sided joint replacements and those who are extremely obese. This law also affects patients who are not on Medicare. Home health services for a few days followed by outpatient therapy programs allow most patients to return home after the short hospital stay. Most patients are independent in their home by one week, are able to drive in three weeks, and are able to return to work that requires prolonged standing by six weeks.

Often overlooked in joint replacement surgery is the importance of a positive mental attitude. If you are not certain that the benefits outweigh the risks of surgery or if you are uncomfortable with your hospital or surgeon or you are just too stressed out, it is best to defer total joint surgery. Patients with unexpectedly low levels of function, unusually severe pain, or poor mental health before surgery have measurably worse results after total joint surgery [28].

## 17. Wheelchairs and scooters

Arthritis can be crippling but not everyone can undergo reconstructive surgery. Severe heart or lung disease, cirrhosis of the liver, kidney failure, the use of immunosuppressive medications after organ transplants, AIDS, morbid obesity, alcoholism and intravenous drug abuse are examples of conditions that are associated with such high complication rates that joint replacement may not be recommended.  In these instances a wheelchair can be prescribed.  Should you require a chair, deconditioning will occur and may eventually rob you of you ability to move from the wheelchair to the toilet or bed.  Inability to make transfers can ultimately require nursing home placement.  It is not usually advisable to attempt total joint replacement after more than six months in a wheelchair as the deconditioning may make return to walking impossible and severe contractures at the hips and knees make the surgery much more difficult.  This decision is always individualized.

A wheelchair can be motorized but insurance may only cover this expense if you also have something wrong with your arms that would preclude you from propelling yourself.  Many orthopedic surgeons do not favor scooters as they believe that scooters may lead to deconditioning and obesity.  Orthopedic surgeons view their role as restoring function and eliminating disability rather than replacing your legs with a scooter.  A physical medicine and rehabilitation specialist may be consulted for a prescription to a wheelchair or a scooter.

# Chapter II. Planning for a Joint Replacement

## 1. Laboratory tests

A set of blood tests and an electrocardiogram are usually done 2-4 weeks before surgery so that abnormalities can be further evaluated and treated. Cross-matching of blood may require another trip to the lab: this test ensures that blood matching is perfect for a possible transfusion, but is only good for three days. Even blood that you donated for yourself (autologous blood) is cross-matched to ensure that there weren't any errors in the processing of your blood.

## 2. Medical clearances

Total joint surgery is an elective operation. There is no reason to take chances with medical problems that can make your care around the time of surgery unpredictable. Your primary care physician should always be involved in the decision to proceed with joint replacement and can often provide any medical clearances required before surgery.

After cardiac surgery, in the setting of known cardiac disease, if symptoms of cardiac disease or EKG changes of cardiac disease are present, further studies will be required. Since an arthritic joint will not allow you to perform to your full capacity on a tread mill stress test, a chemical stress test is often a starting point in this evaluation. In this test medicines (such as persantine) are injected to dilate the heart's blood vessels. Vessels with plaques in them don't dilate as readilly and this results in the uneven distribution of the radio-tracer that has been injected. Radioactive medicines (such as thallium) are then able to image areas of the heart that have diminished blood supply. The test will make you feel flushed and somewhat uncomfortable. If this test is abnormal and suggests a partial occlusion of a coronary artery that could potentially lead to a heart attack around the time of surgery, a cardiac catheterization may be recommended. This will allow the cardiologist to visualize the anatomy of the arteries that supply the heart muscle and may be accompanied by procedures to open up these arteries, such as angioplasty or a cardiac stent placement.

    Coronary artery disease does not preclude the possibility of joint replacement surgery. Many total joint patients have coronary artery disease to one extent or another. Should you require open heart surgery or stent placement, your cardiologist will tell you when you are safe for total joint surgery afterward. A period of time must pass to allow the vessel lining to mature after stent placement and during that time blood thinners are used to prevent occlusion or rupture of the surrounding plaque. This period may be longer for some stents than others. Medicated stents mature more slowly but are less likely to become occluded. Total joint replacement cannot be performed while you are on blood thinners such as Plavix™ which is often used to prevent occlusion of the coronary artery while this maturation of the stent occurs. Many anesthesiologists will not perform regional anesthesia or spinal or epidural blocks unless you can remain off of Plavix™ for 14 days, and Coumadin™ and aspirin for 7 days. In other instances, coronary artery disease does not require any type of surgery but requires management with medications to reduce the chance of a heart attack around the time of the surgery.

    Heart valve disease is often studied with a 2-D echocardiogram. Significant narrowing of the aortic valve, for instance, may pose grave risks for surgical intervention, requiring changes in fluid management and anesthesia. Valve replacement surgery may require the life-long use of blood thinners and require admission to the hospital for a short time before surgery to allow the use of intravenous blood thinners, such as heparin.

    Severe lung disease poses its own set of risks in joint replacement surgery. A set of tests to measure pulmonary functions, oxygen diffusion, and blood oxygen levels will be performed before surgery in this instance. This will allow us to predict the risk of this surgery, to some extent. More importantly, this testing may predict strategies to improve lung function before and after the surgery. A sleep study may be ordered to check for sleep apnea, which requires special monitoring and treatment after surgery.

Anemia (low red blood cell counts) may require additional testing to check for bleeding ulcers and colon cancers which may make blood thinners unsafe. Occasionally doctors prescribe a hormone that boosts blood counts, human erythropoietin, to minimize the need for transfusion in the setting of an anemia. Tests for vitamin deficiencies and of iron levels are also frequently performed to evaluate anemia. Anemia of chronic disease is common in rheumatoid arthritis and in kidney failure patients.

Blood clotting is tested and additional testing may be required if these tests are abnormal, especially if you have a personal history or a family history of either excessive bleeding or blood clot formation. Both high and low platelet counts require evaluation before surgery.

Urinary tract infections are often discovered and treated before surgery. Kidney and liver function tests may also discover unforeseen risks with this surgery. Kidney and liver failure are assosicated with excessive bleeding and higher infection rates in joint replacement surgery [29,30].

## 3. Avoid cuts on your legs

Small cuts from gardening, from pets, or from shaving your legs may increase risks of infections because they increase the numbers of bacteria on the skin. Don't shave your legs with a razor for at least one week before surgery. Report sores or cuts on your legs to your surgeon before surgery as their presence may require rescheduling your surgery.

## 4. Blood donation and blood boosters

Blood that you donate for yourself (autologous donation) can be stored for 42 days. If blood has to be frozen to preserve it longer, it may be lost from breakage of the bag that it is stored in. The most common way to administer the blood booster erythropoietin is to inject 40,000 units of this hormone 21, 14 and 7 days before surgery and again on the day after surgery in average sized patients.

## 5. Stop taking blood thinners

Blood thinners can make joint replacement surgery unsafe. Patients taking warfarin (Coumadin™) prior to surgery should stop this mediation one week before surgery, or as instructed by your surgeon. Plavix™, a medication used to treat coronary artery disease and peripheral vascular disease, has been associated with significant increases in bleeding at the time of surgery. Plavix™ is hard to reverse, should excessive bleeding occur, although aprotinin has been used for this purpose [31]. This medication should be stopped 2 weeks before surgery. Aspirin and most anti-inflammatory medications must be stopped at least one week before surgery as well. Part of obtaining preoperative clearances is to ensure that cessation of these medications for this period of time prior to surgery, and for a period after surgery, will be relatively safe.

## 6. Stop taking herbal supplements

The "four G's"—ginger, garlic, ginkgo bilboa, and ginseng—have long been known to thin the blood and lead to excessive bleeding during surgery. The list of supplements that can interfere with blood clotting is growing, so the best policy is to cease taking all herbal medicines and dietary supplements for one week before surgery.

7. **Medications to take on the morning of surgery**

One class of blood pressure medications, beta-blockers, protects the heart during surgery so these medications are taken with a sip of water before surgery. Another blood pressure medication, clonidine, has been associated with rebound blood pressure changes when it is stopped, and is also taken on the morning of surgery. Clonidine and beta-blockers also reduce the levels of deleterious stress hormones during surgery. Other blood pressure medications, such as calcium channel blockers, can make your blood pressure unstable during surgery and may be withheld on the morning of surgery. Ask your surgeon and anesthesiologist what to take on the morning of surgery.

Oral diabetic medications are usually not given on the morning of surgery. The surgical team will manage your diabetes around the time of surgery. Taking diabetic medications before surgery can result in dangerously low blood sugars during surgery.

A new trend is to take medications before surgery that will help control the pain after surgery. Celebrex™ does not thin the blood but may help control postoperative pain by decreasing inflammation. These medications are prescribed by some surgeons but not by others.

8. **What to wear to the hospital**

Wear comfortable, loose-fitting clothes, as if you were going for a walk. Zip-up exercise pants and walking shoes are good choices. A percentage of total joint patients will be going to a rehab hospital after surgery instead of returning directly home. The clothes listed above will be good choices should you go to a rehab facility.

## 9. What to bring to the hospital

It is advisable to bring your perscription medications to the hospital with you. They can be used if the hospital does not carry your particular medication. Periods of confusion after surgery can easily lead to misuse of your medications. Therefore, in order to avoid overdoses and potentially harmful drug interactions, many hospitals do not allow patients to self-administer their medications. In order to prevent these potential problems, and for the sake of keeping an accurate log of medicine usage nurses generally administer medications.

It is never good to be secretive about your medicines, even tranquilizers and other medicines for psychiatric illness. If your doctors and nurses don't know what you are taking, potentially fatal drug interactions and overdoses can occur.

Bring your own toothbrush, hair care products and makeup. Include anything else that will help you pass the time in the hospital without being too bored.

Bring any living will or advanced directive that you have and wish the hospital and your providers to observe.

## 10. Meeting with your anesthesiologist

General anesthesia renders a patient unconscious for their surgery but, upon full awakening, the patient is aware of the pain from the surgery. Neuraxial anesthesia includes spinal anesthesia and epidural anesthesia, each of which make the legs numb and weak. Numbness is preferable to pain. Patient satisfaction is higher with spinal and epidural anesthesia than with general anesthesia. Patients can still be sedated or unconscious with epidural or spinal anesthesia. Paralysis from bleeding, the most dreaded complication of epidural anesthesia, is very rare, even with post-operative use of the blood thinner, warfarin. The largest study showed no epidural bleeding associated with neurologic injury in nearly 13,000 patients with epidural catheters for up to 48 hours while on warfarin after surgery [146].

One review of 141 older articles comparing epidural anesthesia to general anesthesia found reductions in heart problems, blood clots, lung problems and death rates with epidural anesthesia [32]. Modern improvements in surgical techniques and postoperative care have led to more equivalent overall safety profiles for general and epidural anesthesia. Epidural anesthesia does seem to provide better pain relief, fewer intestinal problems, less narcotic related confusion, faster rehabilitation and possibly fewer blood clots than seen with the use of general anesthesia for total joint patients.

Many other forms of regional anesthesia are available, such as femoral nerve blocks, lumbosacral plexus blocks, and sciatic nerve blocks. What they have in common is the administration of medications that block pain transmission from nerves that reside outside of the spinal canal. Catheters can be placed near the nerves to obtain many hours of relief but "single shot" blocks are not as durable. Nerve injury is rare and low blood pressure is less common than with epidural anesthesia.

Some insurance companies such as Medicare pay your anesthesiologist very little for managing continuous epidural anesthesia after surgery. This may create hesitancy on the part of some anesthesiologists to invest the time required to monitor pain control and blood pressure changes in patients with epidural catheters after their surgery. Don't be afraid to discuss economic issues as well as safety issues with your anesthesiologist.

Remember to tell your anesthesia provider if your previous providers ever had difficulty with the use of breathing tubes during anesthesia, and whether you or a family member have ever had high fever (malignant hyperthermia) from anesthetic agents. Inform them if you have allergies to medications or to latex. Latex allergies require special preparation of the operating room the night before surgery.

**Table 3. Anesthesia Questionnaire**

| | |
|---|---|
| Name: | Age / Date of Birth: |
| Allergies to medications / type of reaction | |
| Allergic to latex / type of reaction | |
| Medical problems | Heart: chest pain, irregular rhythm, murmur, cardiac surgery, stents<br>Lungs: asthma, emphysema, sleep apnea, severe snoring, oxygen use, smoking<br>GI: ulcers, cirrhosis, hepatitis, more than two alcololic beverages daily<br>Nervous system: seizures, stroke, surgery<br>Bleeding problems<br>Other |
| Problems with anesthesia / type of problem | Problems with breathing tube / small jaw / large tongue<br>Difficulty with spinal anesthesia<br>High temperature / malignant hyperthermia<br>Severe nausea / vomiting<br>Family history of problems / type |
| Medicines used / dose / times taken | |
| Medications taken on day of surgery | |
| Prior spine surgery | |
| Blood thinners / when stopped | Coumadin / warfarin<br>Plavix<br>Aspirin<br>Anti-inflammatory medicines<br>Other |
| Use of prednisone or steroids | |
| Use of recreational drugs | |

| | |
|---|---|
| Height / weight: | |
| Preoperative clearances / doctor's name | Cardiology<br>Pulmonary<br>Primary care physician |
| Last food / drink before surgery | Solid food / time<br>Liquids / time |
| Type of anesthesia requested (circle all that you are interested in) | Spinal / epidural / general / regional |
| Special blood transfusion information | Banked blood OK if necessary<br>Autologous donation<br>Do not transfuse me even to save my life |

## 11. The use of immunosuppressive medications around the time of surgery

Medicines that suppress the immune system are often required to prevent rejection of transplanted organs and to treat severe autoimmune disorders. Some of these medications cannot be safely stopped before surgery. Azathioprine (Imuran™) is used in many patients who have had kidney or liver transplantation and cannot be safely stopped for any length of time before surgery due to the risk of rejection of the transplanted organ. One study documented infections in 19% of patients with organ transplants after undergoing joint replacement surgery [33]. Because these medications interfere with the ways that the immune system responds to infection, the usual signs of infection (redness, severe pain and cloudy wound drainage) may be absent in patients on Imuran™. There was an average delay of one year in making the diagnosis of a total joint infection in this study.

For rheumatologic disorders, immunosuppressive medication use around the time of surgery is more complicated and may require consultation with your rheumatologist before surgery. Prednisone is continued around the time of surgery and supplemented with intravenous medication to prevent stress-related circulatory collapse (Addison's disease) that can occur from the suppression of your adrenal glands after long-term use of this medication.

Tumor necrosis factor is a naturally occurring protein that causes inflammation. An excessive amount of tumor necrosis factor is a cause of joint destruction in some diseases. Anti-tumor necrosis factor therapies such as etanercept (Enbrel™), infliximade (Remicade™), and adalimumab (Humira™) carry the risk of infection. There is limited information on the magnitude of these risks to total joint patients. Because cessation of these medications can lead to a flare of the underlying disease (rheumatoid arthritis, lupus, Crohn's disease, or psoriasis), the decision to stop the medicines to reduce infection risks is complicated. Usually symptoms of the inflammatory disease slowly reemerge over one month when these medications are discontinued. These medications usually only have to be stopped for one week before surgery. At the time of this writing, there is one human study that has shown an increase in infections for patients taking these biological medications with orthopedic surgery [34]. More study will be needed to refine our recommendations on the use of these medications.

Table 8 in the appendix is an adaptation of the advice provided by rheumatologists regarding the use of these medications around the time of surgery [35]. The recommendation for withholding anti-inflammatory medications in patients with osteoarthritis, and for the treatment of patients on steroids such as prednisone, is widely accepted. However, the safety of medications that alter the immune system in patients with rheumatoid arthritis, lupus, psoriasis and Crohn's disease when taken around the time of surgery is still uncertain. These diseases are associated with higher infection rates in the absence of medications that suppress the immune system, so it is hard to discern whether it is the disease or the immunosuppressive medication that is the culprit when an infection occurs. The recommendations about withholding these medications around the time of surgery are based on the rheumatologists' more conservative estimates. Without more study, these recommendations are only educated guesses. The authors of this study point out that it may be safe to continue TNA-ase inhibitors around the time of surgery and that there is not enough information available to your physicians to be certain of their recommendations about the use of this class of medications around the time of joint replacement surgery [35].

Medications taken to prevent the rejection of transplanted organs cannot be safely stopped for more than three days. Infection rates are higher in this situation [33].

# Chapter III. Blood Use in Total Joint Replacement

Stated simply, about one in five knee replacement patients and one in three hip replacement patients will require blood transfusion to prevent symptoms from significant anemia or, very rarely, for life-threatening blood loss during or after one of these surgeries. Of every eleven patients who donate blood for themselves before total joint surgery, one patient will also require blood from a donor.

While some of the blood is lost during the course of surgery, more is lost into the soft tissues or into surgical drains after surgery. Although it often shows up later as bruising, the blood lost in the soft tissues after surgery is called "concealed blood loss." It is much more common to need a blood transfusion in the days following surgery than it is to require blood during joint replacement surgery.

Blood transfusion is the replacement of red blood cells which contain hemoglobin, the oxygen carrying molecule. Transfusion may be required in patients undergoing joint replacement surgery. Transfusions are more frequently required in the elderly, in people with preoperative anemia, and in those who use injectable blood thinners after surgery[36]. In one large study, 18% of knee replacement patients and 32% of hip replacement patients who did not donate blood for themselves before surgery required donor transfusions after surgery[37]. In this study 9% of the patients who did donate blood for themselves before surgery still required additional donor blood.

The use of a blood booster, which is a naturally occurring hormone that stimulates the bone marrow to produce more red blood cells, has been shown to decrease transfusion rates. This hormone, which is normally produced in our kidneys, is called "human erythropoietin" and can now be produced as an injectable medication by genetic engineers.

In another study of total joint patients, the selective use of blood boosters (erythropoietin) combined with allowing lower blood counts before transfusion (hemoglobin of 7.0), demonstrated a decrease in the transfusion rates to 1.4% for total knee replacements and 2.8% for total hip replacements [38]. The variable data from these studies reflects a wide disparity among various populations of patients and preferences among various surgeons for the utilization of blood products. Patients with heart disease, lung disease, and other diseases associated with aging will not tolerate low blood counts, and many surgeons would be uncomfortable if they did not transfuse patients with blood counts as low as those tolerated in this study.

The surgeon can roughly calculate how low your blood count will be after surgery. On average about 30% of the body's oxygen carrying molecule, hemoglobin is lost around the time of joint replacement. Hemoglobin levels drop on average 4.0 g/dl for a hip replacement and 3.8 g/dl for a knee replacement [38]. If your surgeon checks your hemoglobin level with a lab test before surgery, your need for blood boosters can be calculated with some accuracy before your operation. For example, if you are somewhat anemic and your hemoglobin level is 10.5 before surgery, and is expected to drop by 4, your hemoglobin level would be only 6.5 after surgery. This level would most likely result in a transfusion. If your predicted post-operative blood count is too low, blood boosters can be prescribed and can be used daily for several days or weekly for three weeks prior to surgery. This prediction of a patient's hemoglobin level is only an estimate. For instance, a hip revision (redo) surgery is generally associated with more blood loss, and thus the preoperative hemoglobin levels may not accurately predict postoperative blood counts for patients undergoing this surgery.

Donating a unit of blood results in a lower hemoglobin level, because not all of the blood is replaced in the month between the donation and the surgery. It was shown that the average decrease in hemoglobin levels was from 14.0 to 12.6 [39]. Since the high risk group for transfusions includes patients with hemoglobin levels below 13, this meant that self donation changed this patient population from 26.2% in the high risk group to 55.7% in the high risk group [39]. The transfusion of blood donated before surgery is more likely by virtue of the donation itself because the blood is not fully replaced by the time the donated unit expires. The use of both erythropoietin and autologous blood (self-donation) has been shown to be more effective that either blood conserving modality is by itself [40].

Preoperative use of erythropoietin is FDA approved for routine use in anemic patients undergoing total joint surgery. Although its use has been shown to decrease transfusion rates in non-anemic patients undergoing total joint surgery, some insurance companies will not pay for these expensive injections unless anemia is present before surgery. Erythropoietin injections are most effective in patients with hemoglobin levels in the range of 10-13. The most common way to prescribe this medication is to inject a small amount (40,000 units) of erythropoietin into the fatty tissue on days 21, 14, and 7 before surgery, and once more on the day after surgery. Blood counts are checked at the time of each injection to ensure that counts are not getting too high as very high counts have been associated with clotting, strokes and heart attacks. The use of erythropoietin is associated with heart failure, death, heart attacks and stroke in dialysis patients whose hemoglobins are allowed to rise above 12 [142].

If you are found to have a more significant anemia in laboratory testing before your total joint replacement, your surgery may be delayed while tests are performed to ensure that you don't have a disease that could compromise the results of your total joint surgery. It is not uncommon to discover that a patient scheduled for total joint surgery has a bleeding ulcer, a cancer of the colon, or a cancer of the bone marrow, such as multiple myeloma. In such cases additional treatment may be required before joint replacement is safe. Patients with rheumatoid arthritis often have an anemia of chronic disease.

One special consideration is that of Jehovah's Witnesses who have a strong opposition to the use of blood products, to the point of refusing life saving transfusions. A surgeon who agrees to operate on a patient who is a Jehovah's Witness has made a commitment to that patient not to use blood products, even to save the patient's life. Many patients who are Jehovah's Witnesses choose to use erythropoietin blood boosters such as EPO™, Epogen™, or Procrit™. This injectable medication will minimize anemia after total joint surgery. While these medicines are not blood products, they are packaged with small amounts of albumin, which is a blood product, to prevent adhesion of the drug to the vial that it is stored in. The Watchtower Society has declared the use of these blood boosters a matter of conscience for each individual [41]. There is may be risks of death, heart attacks, strokes, and heart failure with this medication when it is used to elevate blood counts [142].

Orthopedic surgeons have tried to recapture the blood cells lost during surgery with cell saver devices and soon after surgery with autotransfusion drains, which recover the blood lost after surgery. The use of these devices is controversial, and may not be used by your surgeon unless there is an expectation of high blood loss. The use of cell-saver equipment during surgery typically requires that at least 700ml (1¼ pints) of blood is lost before any blood can be returned to the patient. Blood loss during surgery is often too low to get any blood back from the cell saver except in extensive revision (redo) surgeries. Drains that return the blood lost after surgery may not be of any definite benefit, based on an analysis of the scientific work done on this subject to date [42].

Other strategies exist that can prevent blood loss.

One such strategy, called hemodilution, is a technique that dilutes the blood by removing two pints of blood immediately before surgery and replacing it with fluids. This blood can then be returned to the patient during or after surgery. However, hemodilution could not be shown to be better than self-donation in one study of total hip replacement patients [43]. Hypotensive anesthesia can also decrease blood loss by lowering the patient's blood pressure. With lower blood pressures, the blood loss is decreased. Both of these strategies are unsafe for patients with cardiac conditions. Your anesthesiologist can inform you about whether he or she believes that the use of these special techniques is safe.

Two medications that affect blood clotting have been shown to decrease blood loss from major orthopedic operations [44]. Aprotinin is a drug derived from the lung tissue of cattle, and can cause severe allergic reactions. Tranexamic acid is a drug that is produced chemically rather than derived from human or animal cells. Both have been shown to decrease bleeding without increasing the risks of blood clots in cardiac surgery. These medications are FDA approved for use in heart surgery but more study is required before they can be recommended for use in total joint patients who may have higher risks of blood clots than patients undergoing heart surgery.

No matter what strategies are used to minimize blood loss, you may require a transfusion of blood to prevent stress on your heart from anemia and to allow you to get out of bed after surgery without symptoms from anemia. The risks of transfusion have been greatly decreased with better testing of donor blood. The donors are no longer paid for blood donations. This eliminated the temptation for drug addicts to donate blood for money. It was a major step in eliminating disease transmission.

According to the American Cancer Society the risks per pint of blood are as follows: HIV, the virus that can cause AIDS (1 in 2,135,000), hepatitis B virus (1 in 205,000), and hepatitis C virus (1 in 935,000) [45]. Many risks that we take everyday are greater than the risks of a transfusion [46]. The National Safety Council reports that the lifetime odds of hit and killed by lightening are 1 in 56,439 [47]. In other words, you are 38 times more likely to die from a lightening strike in your lifetime than you are to get the virus that causes AIDS from the transfusion of one pint of blood.

About one in 5,000 transfusion recipients has a transfusion-related lung injury (TRALI). This can be fatal in 5-25% of cases [45]. There are a variety of diseases, some more severe than others, that can be transmitted from a transfusion. Lung injury and rare, potentially fatal allergic reactions and hemolytic reactions (breakage of blood cells) caused by transfusions pose a greater risk to patients than do HIV/AIDS and fatal hepatitis. Allergic reactions to blood and fevers from transfusions can be minimized by using donor blood with the white blood cells removed.

The greatest risk to total joint patients may be from a transient suppression of the immune system after transfusion, called "immunomodulation." We do not know how great this risk is, but one study showed that the infection rate of patients who received banked blood was about double (7% versus 3%) the infection rate of patients who did not receive a blood transfusion [37]. Bladder infections were especially common after transfusion of donor blood. An alternative explanation for the increased infection rates in patients receiving blood in this study is the possibility that patients requiring transfusions are less healthy and are therefore naturally more susceptible to infection. A second study [48], however, showed depressions in several markers of immune function in total hip patients after transfusion. These tests did not become abnormal in total hip patients who were not transfused. More scientific work will be required to clarify this risk. For more statistics on the risks associated with blood transfusion refer to Table 9 in the Appendix.

Great care is taken in the blood banking system to

avoid problems with banked donor blood and self donations. Likewise, great care is taken by the surgeon and the anesthesia provider to prevent excessive blood loss during surgery. There is always a small chance of injury or death from excessive blood loss and a small chance of illness or death from transfusions after total joint surgery. These risks must be weighed against the benefits of this type of surgery.

Donor blood from family members is statistically less safe than banked blood, and it is neither recommended nor is it always covered by insurance companies. Friends and family members may not want to disclose issues relating to sexual promiscuity and previous history of drug abuse. Immune reactions to blood from family members can occur, especially in patients who have weak immune systems. This is called "graft versus host disease," and is the result of white cells from genetically similar blood attacking tissues of the transfusion recipient.

Some surgeons recommend self donation (autologous donation) of blood and others do not. Still other surgeons use self donation selectively. Self donation does not eliminate the possibility that you will require a transfusion of banked blood from a random donor but it decreases this risk. Because nearly half (45%) of the self donated blood is wasted [49], and because self donation is costly, more and more surgeons do not routinely recommend self donations before total joint replacement surgery. Patients who have high blood counts have little risk of needing a transfusion. If the lab showed the blood concentration of hemoglobin to be over 15 g/dl, no patients required transfusion in one study [50]. If the hemoglobin was over 12 g/dl, no patients under 65 years of age required transfusion in this study. Revision (redo) surgery and replacing more than one joint in the same surgery are special situations that may warrant self donation for selective patients.

Not all patients are able to donate blood for themselves. Blood banks often turn away patients with heart and lung problems or poorly controlled blood pressure. Poor vein access often precludes self donation. Anemia is another cause of rejection for self donation. After donation, not all of the blood is remanufactured by the bone marrow before surgery. Inability to donate blood does not require patients to forgo surgery given the safety of banked blood and the limited utility of self-donations.

Even self donated blood is associated with some risks, such as contamination from bacteria (spoilage) or rare clerical errors resulting in the wrong unit of blood being stored.

# Chapter IV. Details of Total Knee Replacement

## 1. What is knee replacement?

A knee replacement is a joint resurfacing operation performed in cases of severe arthritis. Traditionally, the skin incision is about 7" long and is developed into an incision around the inside of the kneecap. Ligaments are balanced to correct the alignment of the knee. The outer edges of the bone are removed and resurfaced with metal or ceramic and plastic (polyethylene). This resurfacing is similar to a crown on a tooth. A very accurate set of jigs is used to guide the surgeon's tools, and precisely cap the surfaces of the knee joint. The kneecap (patella) is retained, but a new plastic surface is often placed on its worn out undersurface. The total knee components are often held in place with an adhesive polymer (polymethyl methacrylate), commonly referred to as bone cement. The alignment of the knee is reestablished so that the previously crooked knee will now be in line with the hip and ankle. The usual angle required to align the knee in this way is 5-7 degrees knock-kneed (valgus). The skin is often closed with staples, which are removed about 10 days after surgery. Motion is restored in surgery but must be maintained in therapy. Short hospital stays and avoidance of post-operative nursing home care are prudent to prevent infections from other patients who may have contagious diseases.

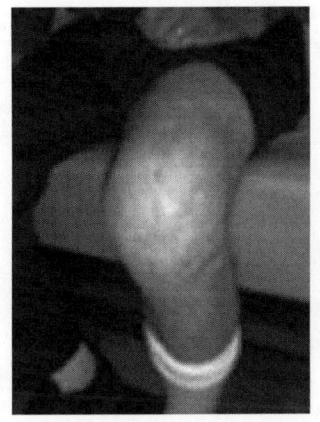

Patients undergoing knee replacement who also have hip arthritis present an additional problem: a contracture around an arthritic hip joint also makes it impossible to walk with the knee straight and will often lead to a knee contracture after knee replacement. A contracture refers to the inability to fully straighten the knee and results in an abnormal gait. We typically prefer to address severe hip arthritis surgically before we replace the knee. Additionally, hip arthritis can cause referred pain to the knee. We are more likely to improve knee pain with hip surgery than we are to improve hip pain with knee surgery.

- **Bilateral knee replacement**

Replacing both knees at the same time is called a bilateral knee replacement. It is theoretically possible that bilateral knee replacement may produce a better result when both knees have flexion contractures as it is impossible to walk on one bent knee while fully extending the other. Try it. Other benefits include avoiding a second hospitalization and avoiding a second period of physical therapy. Bilateral procedures save your insurance company money and convalescence in a rehabilitation hospital is approved by insurance companies and by Medicare more frequently after bilateral procedures. Staggered bilateral procedures involve performing the second knee replacement 4-7 days after the first knee is replaced. Staggered procedures were first proposed after Medicare began paying hospitals and physicians only 50% of their fees for the second knee in 1992. This procedure has been shown in one study to be safer than doing the knees one at a time and to be safer than replacing both knees in the same operation [51].

Several risks are increased with bilateral procedures. Transfusion of banked blood, even after self-donation, is more common if both knees are replaced at the same time. Transfusion is required in 32% of one-sided replacements and in 57% of both-sided replacements [37]. The oxygen carrying molecule in blood cells, hemoglobin, decreased by 5.42 g/dl in bilateral replacement and 3.85 with one-sided replacement in one study [52]. It is associated with a slightly higher death rate than the death rate of one-sided knee replacements [58]. A deep infection that spreads to both knees after a bilateral procedure and requires removal of the implants often produces a worse outcome than occurs when only one knee becomes infected [53]. People under 80 years old are more likely to have gastrointestinal problems with bilateral knee replacement and people over 80 are more likely to have pulmonary problems, such as pneumonia with bilateral knee replacement [27].

In some instances there are overriding social circumstances that make a bilateral procedure preferable to a patient. For instance, some patients may not be able to take sick leave from work for two separate total joint replacements. However, the social and economic benefits must be weighed against the increased risks of a bilateral procedure. Also, both your surgeon and your hospital are only reimbursed 50% as much for the second knee in a bilateral procedure and this may create some reluctance to offer this type of surgery.

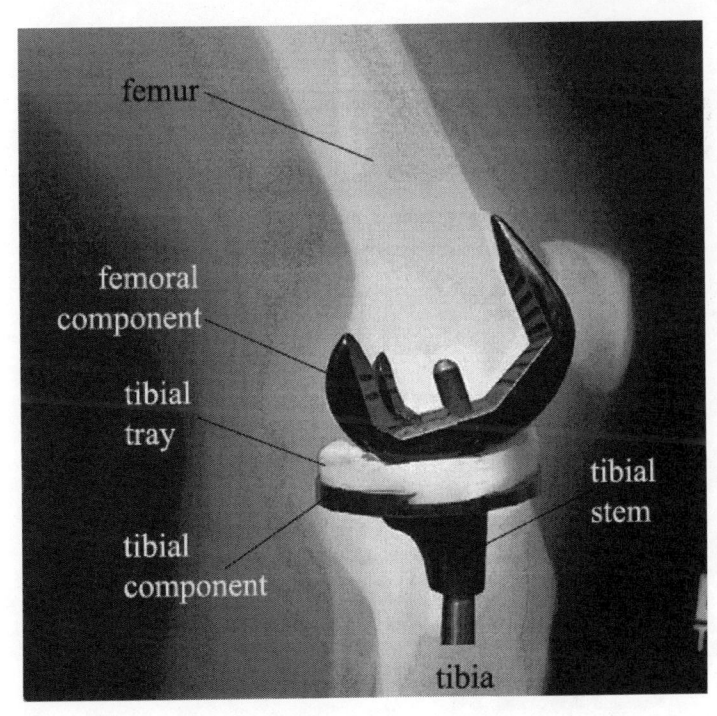

# Pictorial Guide to Total Knee Replacement

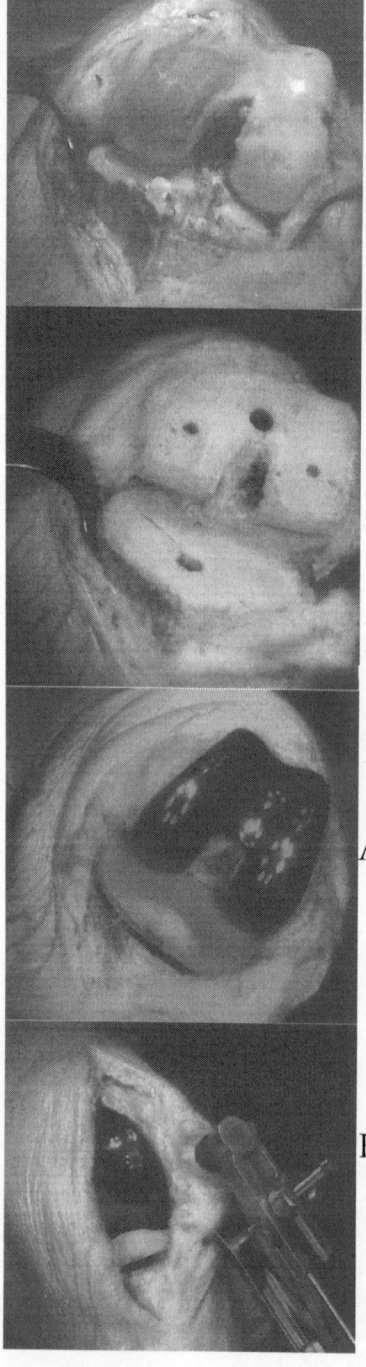

Significant loss of cartilage evident before knee is replaced.

Edges of bone are removed to match the shape of the implants and ligaments are balanced to restore alignment and stability.

Components are glued into place. An oxidized zirconium implant was used in this instance.

Kneecap (patella) is resurfaced with polyethylene.

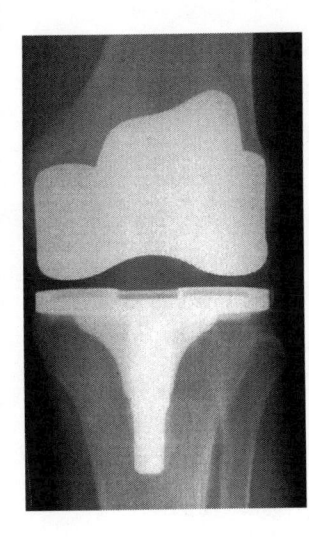

Postoperative total knee replacement x-rays: AP view (left) and lateral view (right).

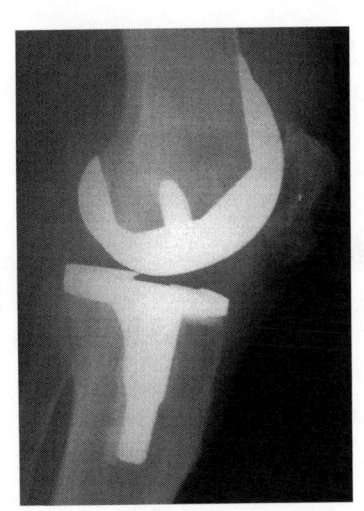

## 2. Risks of knee replacement

Table 4. Major Risks of Knee Replacement

| Selected 90 Day Risks of Knee Replacement in Medicare Patients Who Underwent Surgery in the Year 2000 [1] | Primary (First Surgery) in 124,986 Patients | Revision (Redo Surgery) in 11,726 Patients |
|---|---|---|
| Death | 0.7% | 1.1% |
| Readmission for additional knee surgery | 0.9% | 4.7% |
| Pulmonary Embolism | 0.8% | 0.5% |
| Wound infection | 0.4% | 1.8% |
| Heart Attack (MI) | 0.8% | 1.0% |
| Pneumonia requiring hospitalization | 1.4% | 1.4% |
| Knee manipulation with anesthesia | 1.6% | 1.3% |

The following is a more detailed list of complications (A through Z), although no list can be complete.

A. Loosening and wear: if you live long enough, your knee components will likely wear and become loose or your knee may become less stable from the affects of the inflammation caused by wear particles. Your knee may have to be revised (redone). The durability of the result depends on the technology of the implant, the quality of the surgery, and your activity level.
B. Continued pain: about 5% of all patients have sufficient pain after knee replacement to warrant taking medications. Rarely is pain more severe than the pain before surgery.
C. Infection: In a study of all of primary knee replacements done in Medicare patients the risk of infection was only 0.4% in the 90 days following surgery [1]. Infection rates differ among surgeons and among hospitals. Some surgeons and some hospitals take on higher risk patients and have higher infection rates and some institutions are more honest than others in their reporting of infections. The largest series reported showed deep infection in 2.0% of first time knee replacements and in 5.6% of revision knee replacements [54]. This may reflect the longer follow-up, more honest reporting, and more complex coexisting medical problems treated at that institution. Risks for infection include prior infection, malnutrition, use of medicines that suppress the immune system, rheumatoid arthritis, diabetes, liver or kidney failure, long operations, and poor operating room discipline. A deep infection will require additional surgery – usually removal of the components and placement of an antibiotic spacer or temporary knee prosthesis, followed by replacement of new implants. The surgeon and assistant often wear special exhaust suits and the patient is given antibiotics before surgery, both of which may help prevent infection.

Infections can occur late – months or years after the surgery. It is usually recommended that you take a dose of antibiotics before dental work or urological surgery to prevent late infections that can occur from the bacteria that travel through the blood stream from these procedures. Incurable infections require fusion and, if infection is associated with gangrene, amputation may be required if all else fails.

D. Blood clots (deep venous thrombosis, abbreviated "DVT") often initiate during total joint surgery, but with the use of blood thinners, the severity of these problems can be greatly reduced. DVT can lead to chronic swelling (postphlebitic syndrome) and blood clots to the lungs (pulmonary embolus), which can be fatal. Risks for DVT include: previous blood clots, smoking, the use of estrogen, the use of the medication Tamoxifen™ (used to treat breast cancer), the use of the medication Evista™ (used to treat osteoporosis), old age, obesity, air travel, and being sedentary. You should quit smoking and stop medicines associated with clots one week before surgery. You may restart essential medications a few days after the blood is thinned. The risk of a blood clot traveling to the lungs (pulmonary embolism) is 0.8% [1]. The mean time to the discovery of a clot after total knee surgery is seven days but 42% of blood clots were discovered after discharge from the hospital [55]. With injectable blood thinners, no benefit could be shown for prolonged (30-42 days) of treatment, however [56]. This may not hold true for oral blood thinners such as warfarin (Coumadin™). Modalities shown to be effective in preventing DVT and pulmonary emboli are mechanical (compression stockings) or pharmacological (warfarin, low-molecular-weight heparin, or fundaparinux) [57]. Additional information on this complication can be found in Chapter VI and in the appendix.

E. There is always a patch of numbness on the outside of the incision due to the incision. Neuromas (nerve overgrowths) can form from the interruption of a skin nerve and be painful. However, most patients have only a patch of painless numbness.
F. There is often a click or subtly different feel in the artificial knee. Your knee's new surface is harder than the surface that you were born with. The clicks in the knee represent normal play in the ligaments in most instances. Painful "clunks" in the knee, where the kneecap seems to jump, may require additional surgery.
G. Unsightly scars usually occur when prior scars must be incorporated into your incision and in patients who have a history of excessive scar tissue formation (keloid). Severe problems with wound healing are possible. Consultation with a plastic surgeon may be recommended when there is any reasonable likelihood of these problems. Pigmentation of the incision line is customary for the first 1-2 years and can be minimized by applying sun screen before sun exposure. Aloe Vera, Vitamin E, or silicone strips may improve the appearance of the scar. Lotions and oils should not be applied until the scar is fully healed, generally one week after staple removal. Loss of blood supply to the skin can necessitate muscle and skin grafts.
H. Loss of motion: both you and your surgeon share responsibility for this outcome. It is the surgeon's job to obtain the motion in the operating room and your job to maintain the motion in therapy after surgery. Manipulation of the knee under anesthesia is required in 1.6% of total knee patients to improve motion [1].
I. Transfusions are required in 18% of patients undergoing knee replacement [37]. Transfusions are more commonly required in elderly patients, patients with preoperative anemia, and in patients who use injectable blood thinners after surgery [36]. Self donation of blood before surgery

still leaves a 6% risk for transfusion of donor blood after surgery [37]. Complications of transfusions include but aren't limited to: AIDS (1 in 2,135,000), hepatitis B (1 in 205,000), hepatitis C (1 in 935,000), lung injury (1 in 5,000), potentially fatal transfusion reactions, and fever or chills. For a more complete list, refer to the Table 9 in the Appendix and to Chapter 3.

J. Blood or fluid in the knee & drainage: especially when the blood is thinned from blood thinners, bleeding into the knee and drainage may occur. This will rarely require additional surgery but may slow down therapy.

K. Fractures of bones or ruptures of tendons around the knee are rare but would require additional surgery in many instances. Tendon rupture around the kneecap (patellar ligament) can be very difficult to treat surgically.

L. Swelling or continued warmth to the knee may persist for months after knee replacement. In most instances this is not a sign of infection.

M. Wound or deep tendon repairs can fail (wound dehiscence), requiring additional surgery to repair and to allow the kneecap to track normally. It is important to protect yourself from falls.

N. The knee will rarely feel unstable or dislocate. This is rare for first time knee replacement and is also uncommon for revision (redo) knee replacement.

O. Malposition of the implants: the surgeon may fail to align the knee correctly, leaving a deformity or preventing the kneecap from tracking normally. This is uncommon in the hands of experienced surgeons.

P. Metal allergy: some metals, usually nickel, may rarely cause swelling and a rash. If you get blisters under your jewelry, discuss this with your surgeon before surgery.

Q. Complications from the medications used around the time of surgery will rarely cause grave problems that are life threatening. Blood thinners may cause bleeding into the incision, from the bowel or kidneys, especially if an ulcer or undetected cancer is present. Blood thinners may also cause bleeding into the muscles or brain if an injury occurs or if there is an undetected vascular malformation. Bleeding, allergic reactions and injuries to organs from medications can be life threatening. Rarely, warfarin (Coumadin™) can cause gangrene and heparin can occasionally decrease platelet counts or rarely cause gangrene.

R. Death occurs during surgery in 0.01% and in 0.21% of patients within 30 days of knee replacement in one large series [58]. Death occurs in 0.7% of Medicare patients within 90 days of knee replacement [35]. Over 90% of these deaths occur in patients with heart and lung problems[58] and preoperative evaluation is imperative for patients with significant health problems or risk factors. We lose 16,500 people from bleeding ulcers caused by arthritis medications each year in the United States [4]. Of the 245,000 patients undergoing knee replacement surgery, we expect to lose about 1,715, based on a death rate of 0.7% [1]. Larger numbers of people die in the United States from bleeding ulcers caused by their arthritis medications than from total joint surgery, although there are many more patients taking these medications than there are patients undergoing total joint surgery.

S. Injury to major nerves usually consists of a foot drop, which is the inability to hold the foot up under your own strength. This occurs about 1% of the time and is most common in patients with rheumatoid arthritis, previous nerve injury or disease and is more common in patients who have a severe valgus (knock-knee) deformity, occurring 3% of the time [23]. Foot drop is more common if nerve blocking procedures are used for pain control but the difference is small and this type of anesthesia is still preferable for most patients. Foot drop resolves without surgery about half of the time [59]. Decompression of the nerve may produce recovery if the foot drop persists [23].

Without recovery from this nerve injury, a brace must be worn or ankle surgery can be considered to stabilize the weak ankle. Foot drop can occur spontaneously in patients not undergoing surgery.

T. Component failure: failure of the materials or design of the implants is very rare although an unforeseen recall can occur with any medical device. Recalls do not usually require removal of the implant from your body.

U. Neurologic changes: periods of confusion or slow thinking called "cognitive dysfunction" may occur after anesthesia, from the surgery itself (possibly from circulating fat elements), and with the use of narcotics and sleeping pills. This is more commonly identified in people who already have memory problems such as Alzheimer's disease. This can result in excessive sleepiness, confusion and agitation, which may be seen for several days after surgery. Stroke can occur around the time of surgery.

V. Complex regional pain syndromes, also referred to as reflex sympathetic dystrophy, occur in 0.8% of total knee replacement patients. This condition is characterized by swelling, stiffness, redness, and hypersensitivity of the skin in patients who do not have infection. It is treated with nerve blocks, such as a lumbar sympathetic block.

W. Transfusion may increase risks of infection (immunomodulation).

X. Bowel obstruction occurs in 1.5% of patients and may require a decompressive colonoscopy to allow the bowel to function and to prevent bowel perforation [60].

Y. The risk of pneumonia from the hospital stay is 1.4%[1] and is expected to be lower in patients with shorter hospital stays.

Z. Other hospital acquired infections such as urinary tract infections and contagious colitis (colon infection from a bacterium called *C. difficile*) are common and should also be minimized by keeping hospital stays short.

## 3. Rehab, results, and follow-up care after knee replacement

The goal of early postoperative care is to regain independence in activities of daily living and return home. In most instances, patients return home dirctly from the hospital in two or three days. If this goal is not met in a timely fashion, a safe environment must be found to prevent infections and excessive costs which occur from prolonged stays in the hospital. Secondary goals are to maintain motion of the knee, to restore strength, to control pain, and to restore the ability to safely negotiate obstacles, such as stairs.

Rehabilitation hospitals are sometimes used to provide additional therapy in an environment which is relatively free from contagious diseases. CMS (Medicare) has debated the merits of this care for several years now and coverage through CMS or your private insurance is not guaranteed for admission to these facilities. Current recommendations for inpatient rehabilitation facility (IRF) admission made by the government accounting office in April, 2005 favor patients who have had both knees replaced at the same time, patients over 85 years of age, and patients who are morbidly obese (body mass index over 50) [61].

Nursing homes and skilled nursing facilities are often considered for patients who cannot return home if they are not accepted by an acute rehabilitation hospital. Unless a special skilled facility is dedicated to orthopedic rehabilitation these facilities are the least preferable choice at the time of discharge from the hospital. These facilities are often full of lower funtioning patients and patients with infections, often from resistant bacteria which are contagious.

- **Postoperative care after knee replacement**

Continuous passive motion machines (CPM) may decrease the oxygen supply to the healing incision soon after surgery and their use may be limited in the first few days after surgery. If wound healing problems are anticipated CPM machines may not be used at all. Many insurance companies follow the lead of Medicare and will not pay for a CPM machine for more than 21 days. Motion machines will increase your motion early but the final result (at one year) is not improved with their use. Some surgeons do not recommend the use of CPM machines and there is wide disparity in how they are used.

Bathing is permitted at any time if a water-proof dressing is in place. Otherwise it should be postponed until two days after the staples are removed. Swimming and the use of spas or hot tubs is best deferred until 3 weeks after surgery. This is also about the time that you may resume driving as long as you are not taking pain pills, you have sufficient motion (5-90°), and you feel that you have regained normal reaction time. You may return to work that does not require prolonged standing as early as 3 weeks and to usual standing and walking duties in 6 weeks if your rehabilitation is on target. If therapy progress is very poor, you will require anesthesia for a knee manipulation to break adhesions and scar tissue formation. This is recommended if you do not have 90° of flexion by 6 weeks after surgery. Less than 2% of knee replacement patients require this treatment [1].

In therapy, we expect that you will be at 5-90° by one month and that you will have 0-125° by 3 months. It is important that you maintain the ability to fully extend the knee as it takes more energy to walk on a partially bent knee. For average height individuals, 100° of flexion will be required to alternate feet when on stairs. If you find that you are losing the ability to fully straighten your knee, do not sleep with pillows behind your knee as the healing process can cause a flexion contracture from the long periods of time that the knee is healing in a bent position. If a contracture is developing, a knee immobilizer or dynamic extension splint may be prescribed for nighttime use. To improve flexion outside of your physical therapy, use a stationary bicycle and try to lower the seat a little each week. Pulling yourself around with your feet on a wheeled chair will increase flexion, as will doing squats or sitting on the stairs one stair above your feet. While many will regain more than 125° of flexion, some will not. In extreme preoperative knee disease, the quadriceps will have developed shortening that will permanently limit the knees ability to flex no matter how diligent you are in therapy. Remember that there is a balance in therapy. Too much therapy will increase inflammation (swelling, redness and warmth) and may lead to a worse result. Too little therapy will lead to weakness and loss of motion from scar tissue formation as well as deconditioning. Patients who struggle with physical therapy will take longer to recover from this surgery.

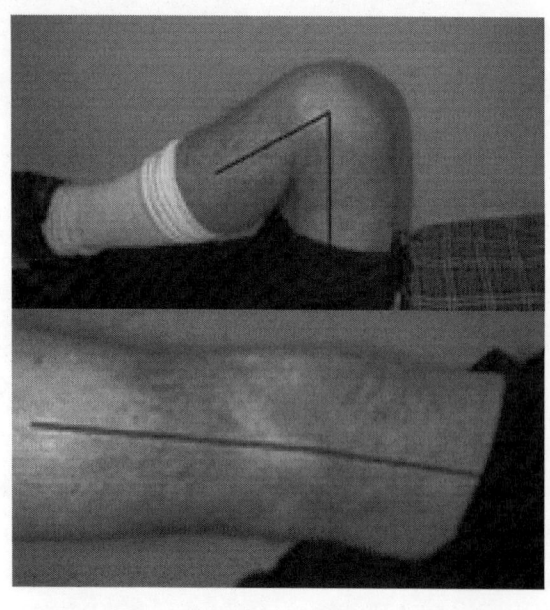

You may be placed on an injectable blood thinner to prevent blood clots for a period of time after surgery. It is simple to learn how to perform these injections on yourself or for a loved one to do this for you. Laboratory monitoring is often not required for the short term use of these medications.

For a period of time after surgery you may be on the oral blood thinner, warfarin (Coumadin™), to prevent blood clot growth. Each week you should go to the laboratory and have a blood test (protime with INR). Either your surgeon or your primary physician will adjust this medication to keep the INR between 2-3. Warfarin will be adjusted around each of the medicines that you usually take but, because it has interactions with most other medications, you should call your surgeon's office or your primary doctor's office, to discuss the use of warfarin with the other medicine. This includes over-the-counter and herbal medications. Occasionally, ulcers and growths in the colon or urinary tract may lead to blood in the stool, black stools or blood in the urine. Should this occur while you are taking warfarin, stop the medicine and seek immediate medical care. Aspirin can also be used to prevent blood clots, but many physicians do not believe that aspirin is effective at preventing clot growth in veins.

If you develop swelling or pain in the calf or thigh, this can be a sign that you have developed a clot in a deep vein (deep venous thrombosis, or "DVT") and you should seek medical care. Should you become short of breath or have chest pain, this can represent a clot in the lung (pulmonary embolus) or a heart attack (myocardial infarction). These complications are life-threatening and you should call 911 immediately if they occur.

- **Resuming activities and long-term results after knee replacement**

    Sports should be deferred for 12 weeks. Many court and field sports and all contact sports are not advised. You should be able to hike, golf, play tennis [62], and bicycle. If you were able to ski before surgery, you can resume skiing at this point. Don't be afraid to be active after your knee replacement – that's why you had your surgery. After a knee replacement you should not try to learn to ski or ride motorcycles due to the risk of fractures. Fractures around the knee joint will be more difficult to fix than they would otherwise have been before your knee replacement.

    Sexual relations can be resumed whenever it is comfortable to do so. Kneeling may result in injury to the skin immediately after surgery, and should be postponed for 6 weeks to allow for adequate healing. Unfortunately, some patients may never be able to kneel comfortably after a knee replacement.

Travel is reasonably safe after joint replacement while the blood is thinned with injectable blood thinners or warfarin. After discontinuing the use of blood thinners there is a period of time that you may be at risk for blood clots in the veins or clots that pass to the lungs (pulmonary emboli, which can be life-threatening). The risk can be decreased to some extent by taking breaks to stretch your legs. Especially common with air travel, blood clots can occur with prolonged sitting. Exercise and avoiding dehydration are recommended but the use of aspirin during travel is no longer recommended by the American College of Chest Surgeons to prevent clots in leg veins during travel. Should you need to travel soon after total joint replacement surgery, you should discuss the use of injectable blood thinners with your physician.

Most studies on knee replacement demonstrate good to excellent results (with minimal pain or no pain, as well as good motion and restoration of function) in 95% of patients. With improvements in surgical technique and engineering, knee replacements have become much more durable. A series of 11,606 total knees performed at the Mayo clinic from 1978-2000, reported that 78% (nearly four out of five) of the replaced knees could be expected to last for 20 years [63]. The surgeons at the Mayo clinic demonstrated good durability of their knee replacements in younger patients with 83% doing well at 10 years compared to 94% of the knees doing well at 10 years in patients over 70 years old. Older patients do not wear out their knee replacements as fast as younger patients do. Implants were slightly less durable when surgeons removed the PCL (posterior cruciate ligament) or didn't cement all components, and more durable in older patients, in women, in patients with rheumatoid arthritis, and in patients with no prior surgery in this study. New developments recently introduced such as the surface ceramic knee and improvements in plastic manufacturing may extend the durability of the replaced knee significantly. As with any new technology, there is a small chance that our new implants will fail to meet the standards of traditionally designed knee implants. For that reason, it may not be wise for older patients to take this risk.

- **Antibiotics for dental work and urologic procedures**
Antibiotics before dental and urologic procedures are recommended. The risk of an infection traveling through the blood steam after a dental procedure is about one in 10,000. Two grams of ampicillin, cephalexin or cephradine is prescribed for use 60 minutes before the procedure for patients not allergic to penicillin-type medications. For patients allergic to penicillin, clindamycin 600mg is substituted. For patients with multiple severe allergies it is probably best not to use antibiotics before these procedures unless there are risks for infection from medications that suppress the immune systems, from other diseases, or from the nature of the dental or urologic procedure.

- **Long-term follow-up with your surgeon**
Every one to two years many surgeons will take X-rays to ensure that the joint is not loosening, not wearing excessively and not causing bone to thin around the implant. You should seek an earlier appointment if there is a new injury, swelling or loss of stability of the joint.

- **Giving the surgeon a grade**
There are a variety of patient questionnaires that are useful, but the tool that surgeons most often used to study their results is a combination of subjective and objective factors, originally called the HSS (Hospital for Special Surgery) knee score. Described by Dr. John Insall in 1976 [64], this tool was used by his institution from that time forward with two later modifications that then were called the KSS (Knee Society Score). The 1993 modification of the Knee Society Score is included in Table 11 in the appendix.

# 4. Technologies and techniques for knee replacement

- **Reduction of backside wear**

    We expect wear from the parts that move on each other after knee replacement, but not from the parts where there should be no motion. With this in mind, improvements in locking these components and polishing of components have greatly reduced this wear. The use of all-polyethylene tibial components and non-modular components eliminate this wear but also eliminate the surgeon's ability to fine tune knee stability at the end of the surgery.

- **Surface ceramics (oxidized zirconium)**

    Traditional ceramics have not gained wide-spread acceptance in knee replacements because of concerns about breakage and the expense of these implants. The use of ceramics in the knee to reduce wear by decreasing scratching and reducing friction is appealing, however. The breakthrough in the use of ceramics in knee replacement is the development of the surface ceramic, Oxinium™. It is as fracture-resistant as metals but it has a thin surface of zirconia, a ceramic. Created through a chemical transformation of the surface where motion occurs against the plastic (polyethylene), Oxinium™ does not delaminate. Wear reductions varying from 42-85% have been demonstrated with this device. It cannot be used without cement to affix it to the bone.

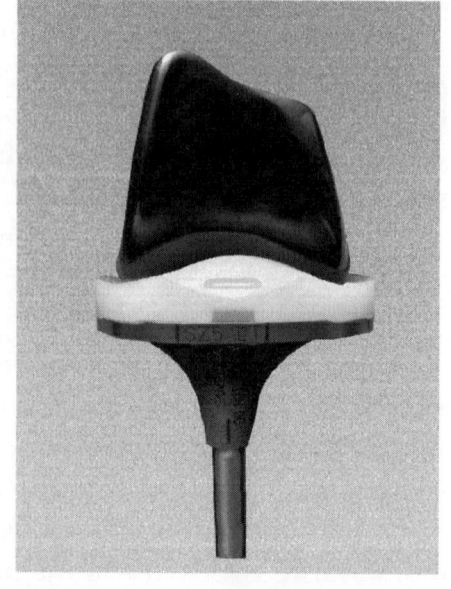

- **Mobile bearing and rotating platform knees**

Although knee surgeons originally felt the rotating platform knee system would be more forgiving, revision surgeons are seeing failures from this device. This device appears to reduce stresses on the cement used for fixation and on the polyethylene insert, suggesting that the durability of this implant might be increased, compared to fixed bearing knees. When the knee's ligaments are not properly balanced, however, the rotating platform can dislocate, requiring revision surgery soon after the initial surgery [65, 66]. The frequency of this devastating problem has ranged from 0-9% [65]. This wide range of tray dislocation rates suggests that either the surgery is more demanding than fixed bearing knee replacement surgery or that the details of the design of the implants have been problematic.

There are two other problems with the rotating platform knee in its current design. The second problem is that their motion has been inferior in some studies but not others. A study from 1997 showed that these devices produced an average of only 108 degrees of flexion compared to fixed bearing knees that resulted in 123-127 degrees of motion [67]. In another series only an average of 102° of flexion was obtained in patients with rotating platform knees[68]. This group still had only 105° of flexion 15 years after their surgery but none of the implants from this group have required revision (redo surgery for wear or loosening), proving that they are durable [69].

The third problem with rotating platform knees is disolving bone (osteolysis). When fixed platform knees wear out, the polyethylene particles are large and usually can't be absorbed by the scavenger cells that contribute to dissolving bone (osteolysis) around implants. In a study comparing dissolving bone in fixed and mobile-bearing knees, osteolysis was seen in 47% of the mobile bearing knees and in 13% of the fixed bearing knees [70]. Increased osteolysis suggests that the motion under the rotating platform may create wear mechanisms like those previously seen in total hip replacements. Osteolysis has been more common in hips than in fixed bearing knees. Reduction of the wear from mobile bearing knees may be possible with the use of alternative bearing surfaces and improved plastic manufacturing similar to those that successfully reduced osteolysis in hip replacement surgery, negating this concern.

The rotating designs for knee replacement have been in use in the U.S. for over 20 years now. There is no consensus that mobile bearing knees available at this time are more durable than the conventional fixed bearing knees. There has been only one randomized study of mobile versus fixed bearing knees to date. This study showed more implant related failures in the mobile knees but otherwise equivalent motion and results at just over three years after implantation [118]. Another study that lends strong support to rotating platform knees involved 116 patients who had one knee replaced with a fixed bearing knee and the other knee replaced with a rotating platform knee. This study, performed in South Korea, showed that the range of motion, failure rates and knee scores were nearly identical 6-8 years after the surgery [119]. With continuing refinements in designs and surgical techniques, rotating platform knees may realize their potential to increase the longevity of the total knee replacement.

- **Total knee replacement without patellar (kneecap) resurfacing**

Although common in the United States, resurfacing the patella is uncommon in some other countries. While some long-term results appear equivalent, many surgeons fear that not resurfacing the kneecap may result in some continued pain from arthritis under the kneecap. Patients with rheumatoid arthritis have been shown to have inferior results when the kneecap is not resurfaced. Certain designs are better if the kneecap is not resurfaced because they match the anatomy of the kneecap better than others. When pain under the kneecap does occur after a knee replacement is done without resurfacing, another operation to add a plastic surface to the underside of the patella may be recommended. Called secondary patellar resurfacing, this surgery does not always alleviate pain in the kneecap.

- **Unicompartmental knee replacement and isolated patellofemoral resurfacing**

Unicompartmental replacement refers to replacement only of the inner or outer compartment of the knee. There are three compartments to the knee: the medial (inside), the lateral (outside) and the patellofemoral (front compartment beneath the kneecap). In unicompartmental arthroplasty it is usually the medial compartment that is replaced, but excellent results with replacing the lateral compartment have also been reported.

Some surgeons do not offer unicompartmental arthroplasty for the same reason that they always resurface the kneecap: they are worried that the non-resurfaced compartment(s) will wear out. Additionally, there is concern that ligament balancing may be more difficult with surgery on only one compartment. Less accurate restoration of alignment in unicompartmental knees than in total knees was found in one study [71]. The third concern is that some of the biochemical changes in the joint fluid affect all three compartments and addressing only one compartment may be inadequate to the task of providing long-term durability. Finally, unicompartmental knee replacement may not adequately improve the loss of motion that can occur with arthritis. Especially in patients who cannot straighten their knee due to a flexion contracture, total knee replacement is the preferred operation.

While the results seen in one study [72] have been as durable as total knee replacement, the patients selected for unicompartmental arthroplasty historically have been older, less obese and less active. It remains to be seen if patients with wear in only the medial compartment will be equally well served with either type of knee replacement, regardless of deformity, age, weight, and activity over the very long term.

Should further long-term studies demonstrate equivalent long-term results for younger, more active patients, unicompartmental arthroplasty looks to be a good option. The knee moves more normally (normal kinematics) if the ACL is intact. Patients who have a total knee replacement on one knee and a unicomartmental knee replacement on the other knee find that the unicompartmental knee feels more natural than the total knee [73]. Small-incision surgery is easier in a "uni" and is truly less invasive than standard total knee arthroplasty.

Isolated replacement of the patellofemoral joint (the anterior compartment) is not widely done in the United States. Since the rotation of the knee joint cannot be manipulated with this proceedure, problems with kneecap tracking are a source of failure and may require other realignment procedures at the time of surgery to prevent an unstable patellofemoral (kneecap) joint.

The use of a mobile unicompartmental implant, called the UniSpacer™, has been associated with a 21% early revision (redo) rate and cannot be recommended [74].

- **Small incision surgery and quadriceps sparing technique**

Believed to speed up rehabilitation, surgery through limited incisions has not yet been shown to influence the long-term result of knee replacement. Many surgeons are uncomfortable with an incision too small to allow perfect visualization of the bone and implant during implantation and cement removal.

Just as in small-incision hip replacement [75], small incision knee replacement can be associated with component malposition [76]. Other surgeons do not report problems with component malposition [77]. Malpositioned components and poor mechanical alignment of the knee have been shown to decrease the durability of a knee replacement.

The use of incisions that do not split the quadriceps tendon are believed by some surgeons to cause less postoperative pain. The early experience has included improved pain control over traditional postoperative care. We are not yet sure if the better early results owe to the new surgical techniques or to the improved pain management. Additional study is underway to discover the benefits and risks of small incision surgery. A multi-center trial of mini-incision versus standard incision knee replacement showed no differences in blood loss, operative time, infection, or knee scores at 12 weeks after surgery but the small incision group had more skin healing problems[78]. The better results ititially reported for "minimally invasive" surgery appear to relate to the improved anesthetic techniques and postoperative pain managment techniques used by the surgeons that developed small incision surgery[144]. You should discus this new protocol with your surgeon and anesthesiologist. It is listed in Table 19 in the appendix.

- **Previous patellectomy (removal of the kneecap)**

Two or three decades ago, surgeons used to remove the kneecap to treat kneecap pain. This is still occasionally required to treat a severe injury. Absence of the kneecap leads to inferior results with knee replacement. Three options exist for knee replacement after patellectomy: grafting an organ donor's kneecap and tendons, accepting the inferior results and replacing the knee without reconstructing a kneecap, or replacement with a specialized implant, the Augmentation Patella™, which has potential to ingrow into the remaining tendon that used to house the kneecap. Ingrowth of this device into a bare quadriceps tendon may be compromised unless there is some bone stock remaining, however [79]. It may be advisable to use posterior stabilized implants after patellectomy [80].

- **Moderately and highly cross-linked polyethylene**

Because wear mechanisms and the forces on the plastic (polyethylene) are different in the hip and the knee, the highly cross-linked polyethylene used for the hip may not provide the same benefits in the knee. One manufacturer of moderately cross-linked polyethylene has been allowed to advertise reduction in delamination by the FDA. Pitting and delamination of the plastic (polyethylene) are the major modes of wear in total knee replacements. Optimal cross-linking for the knee is still being studied and it may be different for the tibial plastic and the plastic used to resurface the kneecap. Cross-linking without melting has been shown to maintain the strength of a brand of polyethylene called x3™ poly, produced by Stryker. The improvements in the chemistry and sterilization techniques have already greatly increased the wear resistance of polyethylene. Additional study will be required to see if cross-linking will provide even more durability or there is any reduction of dissolving bone (osteolysis) around rotating platform knees which may be more susceptible to this disease.

- **Computer navigation**

Navigation systems allow the average orthopedic surgeon to be a bit more accurate with knee replacements. However, both the additional time spent in surgery and the presence of more personnel in the operating room can contribute to increased deep infection rates [54]. Also, the costs of the systems can be substantial to the hospital that may already be losing money on the surgery in their Medicare patients. Navigation may be particularly beneficial for surgeons attempting to use minimally invasive techniques and is under study to see if it will improve long-term surgical results.

- **Retention of the posterior cruciate ligament**

The anterior cruciate ligament (ACL) is removed in total knee surgery with current designs. Removal of the posterior cruciate ligament (PCL) can also be performed with the use of a peg in the polyethylene that provides this stability. Proponents of PCL removal cite improved range of motion and improved ligament balancing especially for knees with a great deal of deformity (contractures, bowed-knees or knock-knees). Those who advocate retaining the PCL believe that it is a secondary stabilizer which may protect the knee from instability in an injury and that there will be less concern about implant failure when the polyethylene post is not used to substitute for this ligament. The PCL is rarely retained in revision surgery. The Journey™ total knee, recently approved by the FDA, uses a plastic peg to substitute for both cruciate ligaments and is reported to have more nomal motion (kinematics) than other designs.

One study showed lower rates of instability with PCL substituting technique in patients with rheumatoid arthritis[81], and one study showed better results with PCL substitution in patients who had previously had their kneecaps removed [80].

The decision about PCL retention must be left to the surgeon. Differences in these techniques appear to be small.

- **Use of all-cemented technique in knee replacement**

All-cemented knee replacement is probably more durable than operations which cement only some of the components [62]. Uncemented metal-backed patella (kneecap) replacement has clearly been associated with higher failure rates with some designs of total knees. Uncemented tibial and femoral components (the major components) are not favored by some surgeons due to the occasional failure of bone ingrowth into the implants. The use of cement is required for fixation of surface ceramic (Oxinium™) to the femur in knee replacement.

- **Use of all-polyethylene tibial components**

There is no backside wear with this technique and the implant is less expensive than implants which use metal backing. This technique does not allow for liner exchange if the plastic wears out, although in one study 25% of revisions with simple liner exchanges failed in less than 5 years [82].

This implant is cost-effective and decreases wear but is more demanding for the surgeon due to the increased difficulty in the removal of excess cement. Also, since the plastic is not easily interchangeable (modular), the surgeon loses the ability to change the tray thickness late in the surgery should soft tissue injuries occur during implantation or cement removal. Because cement removal is more difficult and the stability of the knee can't be fine-tuned late in the case, the use of the all-polyethylene tibial component may work better for very experienced surgeons.

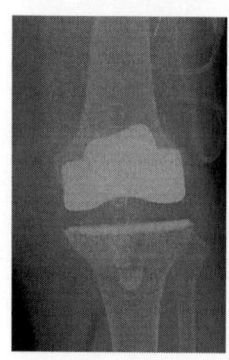

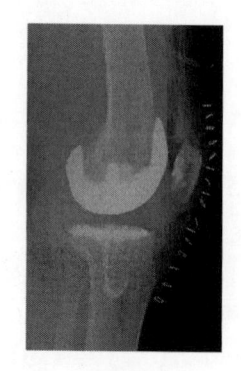

- **High volume surgeons and hospitals**

Complication rates are lower among surgeons and hospitals that perform higher volumes of total joint surgery. For instance, the deep infection rate is 40% lower in the patients of high volume surgeons (0.29% versus 0.55% in the first ninety days after surgery) [83]. The rate of medical complications is 2.9% with high volume surgeons compared to 4.0% with the lowest volume surgeons [40]. This additional margin of safety is not overwhelming and you can find orthopedic surgeons in most areas of the country that will have outstanding track records with most joint replacement procedures.

- **Note and questions for your surgeon:**

# Chapter V. Details of Total Hip Replacement

## 1. What is hip replacement?

If you could choose a joint in your body to wear out, you would choose the hip. The success of total hip replacement is very high, the operation and rehabilitation is less painful than total knee replacement, and hips are extremely durable, with historically good potential to last for more than twenty years.

The hip is replaced with a cup which is pressed into the worn out socket and a stem affixed to a ball which is pressed or cemented into the marrow cavity of the thigh bone (femur). There are a variety of techniques used to affix the implants into the bone and a variety of materials used. There are also a variety of surgical approaches used.

The surgery is accomplished through a 5-9 inch incision in most instances (as short as 2 ½ inches in mini-incision surgery). The operation takes about one to two hours and will sometimes result in enough blood loss to warrant a transfusion. Skin closure is often with staples to reduce the amount of absorbable suture in the wound. If skin closure is with suture and Steri-Strips™, blisters may form at the edges from traction on the skin as swelling occurs. Do not drain them. They are generally left alone, although a cream, such as Silvidine™, may be prescribed if they rupture.

Hip replacement is often performed with cementless fixation, and you may be asked to avoid anti-inflammatory medicines and excessive activity for six weeks after the surgery. Smoking and the use of anti-inflammatory medications may block bone ingrowth into the implants. Excessive activity can lead to small amounts of motion between the implants and your bone and slow down ingrowth. You may be asked to keep some weight off of your hip for six weeks. Elderly patients are often reconstructed in ways that allow full weight bearing on the hip beginning the day after surgery such as using partially cemented (hybrid) hip replacements where the stem, but not the cup, is cemented into place. There is evidence that either cementless or hybrid techniques are preferable to fully cemented hips as most surgeons report inferior long term results with the cemented cups.

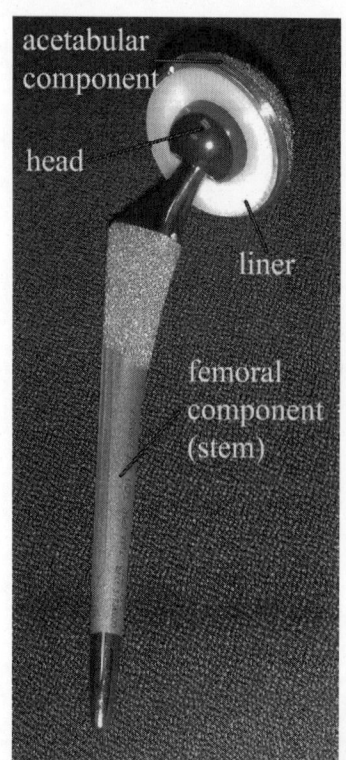

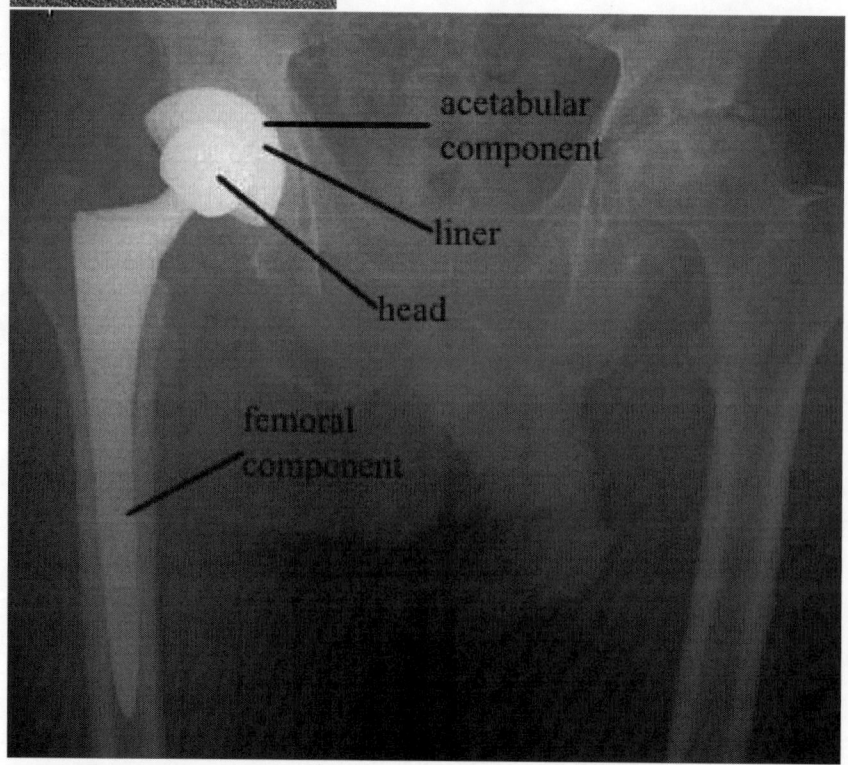

## 2. Risks of hip replacement

### Table 5. Major Risks of Hip Replacement

| Selected 90-Day Risks of Hip Replacement in Medicare Patients Who Underwent Surgery between 7/1/95 and 6/30/96 [2] | Primary (First Surgery in 61,568 Patients) | Revision (Redo Surgery in 13,483 Patients) |
|---|---|---|
| Death | 0.97% | 2.6% |
| Readmission to hospital | 4.6% | 10.0% |
| Pulmonary Embolism | 0.93% | 0.79% |
| Wound infection | 0.24% | 0.95% |
| Hip dislocation | 3.1% | 8.4% |

The following is a more detailed list of complications of hip replacement (A to Z), although no list can be complete.

A. Loosening and wear: if you live long enough, your hip components may wear and become loose or unstable from dissolving bone and joint inflammation caused by wear particles. Your hip replacement may have to be revised (redone). The durability of the result depends on the technology of the implant, the quality of the surgery, your size, and your activity level. New technology, such as ceramic hips, all-metal hips, and highly cross-linked polyethylene may greatly extend the durability of the hip replacement. Increasingly, we are able to do liner exchanges for excessive wear, without replacing the other components. With modern components which are well-fixed (no loosening or severe dissolving bone), this type of revision requires much less surgery. A small percentage of patients with ingrowth hip replacements will fail to ingrow bone into the implants and this will typically require revision surgery. If a tapered, collarless stem is placed, it may slide a bit further down the bone to a more stable position and obtain ingrowth. Migration of the stem down the canal is called "subsidence." If there is significant subsidence it can lead to enough shortening or instability to warrant additional surgery. Subsidence often causes out-toeing due to changes in the rotation of the loose hip prosthesis.

B. Leg lengths may change; especially in revision (redo) surgery. Most commonly, the operated hip is made longer to increase stability of the joint. A shoe lift on the opposite foot is occasionally required. Surgery to correct the problem is rarely required or recommended. Failure of bone ingrowth may lead to shortening. Functional leg length inequality from a tilt of the pelvis (pelvic obliquity) occur in 14% of hip replacements and improves over time[84].

C. The limp may not resolve. Limping can occur if there is continued pain, if the surgeon fails to reestablish the offset of the hip (which balances the muscle forces), or if the muscles around the hip fail to heal (if the surgeon uses certain approaches which detach muscles).

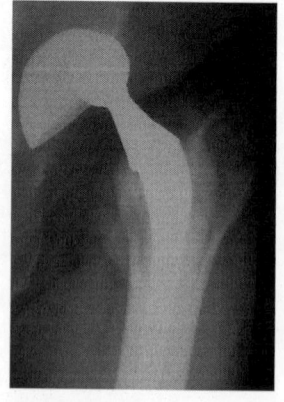

D. Dislocation: the artificial hip may dislocate, especially if it is flexed too far or rotated with the knee pointing in. A dislocation usually results in a trip to the emergency department for anesthesia to relax the muscles so that it can be reduced (re-located) with manipulation. A brace may be recommended for 6 weeks although evidence to support bracing is ambiguous. Additional surgery may be required if the problem is recurrent. However, 50% of early dislocations never recur. There is some variability in the risks of dislocation from surgeon to surgeon depending on the surgical approach used, whether the hip capsule is balanced and repaired, the type of components used, and the position of the components placed by the surgeon. The risk of dislocation is about twice as high in women as it is in men. Patients with dementia, alcoholism and neurological problems are at greater risk because of falls and loss of protective reflexes. This risk of dislocation has been reported to be as low as 0.4% and as high as 11%. In the largest series of patients it was reported as 3.1% for first time hip replacement and 8.4% for revision surgery [2]. Dislocations may increase with time after hip replacement[140]. Recurrent dislocation is a frequent cause for revision surgery.

E. Continued pain: about 2% of all patients have sufficient pain after hip replacement to warrant taking medicines. Rarely is pain more severe than the arthritis pain that made them seek hip replacement in the first place. 98% of patients report no pain or minimal discomfort. Thigh pain from the implant is the most frequent type of complaint but only in an unusual case would additional surgery be recommended for this type of pain which is generally mild.

F. Blood clots (deep venous thrombosis, abbreviated "DVT") often form during total hip surgery, but, with the use of blood thinners, the severity of these problems can be greatly reduced. Pumps worn on the legs may be effective at preventing clots in the pelvis that we can't detect on ultrasound [86]. Clots that form in the pelvis require MRI or CT to detect. They occur in 10.4% of patients undergoing hip replacement and are generally missed on routine ultrasound studies [87]. A DVT can lead to chronic swelling (postphlebitic syndrome) and blood clots to the lungs (pulmonary embolus), which can be fatal. Risks for DVT include: previous blood clots, smoking, the use of estrogen, the use of Tamoxifen™ (used to treat breast cancer) and the use of Evista™ (used to treat osteoporosis). Other risks for DVT are old age, obesity, air travel, and being sedentary. You should quit smoking before total hip surgery and you should stop medicines associated with clots one week before surgery. You may be allowed to resume essential medications after the blood is thinned. Long surgical times and greater blood loss have also been associated with more clot formation. Blood clots that travel to the lungs (pulmonary embolus) occur in 0.9% of hip replacement patients and can be fatal [2]. The average time from hip replacement surgery to discovery of a blood clot is 17 days and 76% of the clots are discovered after discharge from the hospital [55]. Blood thinners should be continued after being discharged from the hospital. The risk of DVT was decreased from 4.3% to 1.4% by continuing injectable blood thinners for 30-42 days[56]. Modalities shown to be effective in preventing DVT and pulmonary emboli can be mechanical (compression stockings or pumps) or pharmacological (warfarin, low-molecular-weight heparin, or fundaparinux) [57].

G. Fracture: the surgeon may create a fracture while press-fitting an ingrowth femoral or acetabular prosthesis. This will usually require placement of a cable but is not expected to compromise the result in most instances. In rare instances a fracture can require significant additional surgery and special equipment to repair. Risk is usually about 1% in a primary hip replacement, and higher in revision surgery.

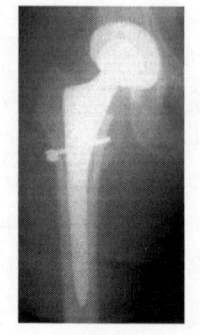

H. Infection: Infections occurred the first 90 days in 0.2% of primary (first time) joint replacements and in 0.95% of revision surgeries in a study of all of the total hip replacements done on Medicare patients in 1996[36]. Infection rates differ among surgeons and among hospitals. The largest series reported showed deep infection in 1.3% of first time hip replacements and 3.2% of revision hip replacements [54]. The institution in this study is famous for taking care of high risk patients and honest reporting of complications. Some infections are identified late and are not reflected in studies that include only the infections discovered in the first 3 months after surgery. Risks for infection include prior infection, malnutrition, use of medicines that suppress the immune system, rheumatoid arthritis, diabetes, liver or kidney failure, and long operations. The surgeon and the assistant often perform the surgery in exhaust suits to decrease infections. Antibiotics are administered just before surgery to help prevent infection. Infections can occur late, months or years after the surgery, so antibiotics are recommended for dental and urological procedures to prevent spread of bacteria through the bloodstream. Deep infections require additional surgery, usually removal of the components with placement of a temporary hip followed by another revision to replace the hip with durable components. If the infection cannot be eradicated, the components are removed and the hip joint is left empty, resulting in variable degrees of pain and a profound limp and patients require a walker in most cases.

I. Loss of motion: the inflammation from surgery may result in the formation of calcium deposits in the soft tissues around the hip (heterotopic ossification, abbreviated "HO"), which can cause the artificial joint to lose motion. If certain risk factors are present, such as previous HO with hip surgery or diseases associated with very large bone spur formation, low dose radiation therapy may be recommended to you to prevent this complication. Low dose radiation (700 cGy) is used to treat patients at risk for severe HO, compared to the high dosage (5000 cGy) used for tumors. Mild HO occurs frequently, but it is rare to have motion limited or to have the result compromised by this process.

J. Unsightly scars: the scar from the incision will gradually become the color of the surrounding skin over a two year period, and is rarely a source of complaint for patients. The fat around the hip may atrophy causing an asymmetrical appearance, and the skin may become discolored from bleeding around the time of surgery, but both occurrences are uncommon.

K. Bleeding & drainage: especially if your blood is thinned from blood thinners, bleeding into the incision, bruising and drainage may occur. This will occasionally require additional surgery.

L. Neurologic injury: Periods of confusion or cognitive dysfunction (slower thinking) may occur after anesthesia and surgery, possibly from circulating fat elements or from the use of narcotics and sleeping pills. This is more commonly identified in people who already have memory problems such as Alzheimer's disease. This can result in excessive sleepiness, confusion, and agitation, and may be seen for several days after surgery with subtle changes persisting for longer periods of time. Strokes can occur around the time of surgery.

M. Death from heart problems, stroke, infection, organ failure, and blood clots (pulmonary emboli) is uncommon (1.0%) [2]. Interestingly, survival rates after recovery from hip replacement surgery appear to be increased for about five years, compared to patients not undergoing this operation [88]. All medical problems that create a high-risk situation should be evaluated by the appropriate specialist before hip replacement surgery.

N. Injury to major nerves, blood vessels, or internal organs are more common with revision (redo) surgery. The most common nerve injury is to the sciatic nerve, causing a foot drop and occasionally severe pain. This occurs in 0.17% of patients, more commonly in patients with hip dysplasia, posttraumatic arthritis, and with a posterior approach to the hip, lengthening of the extremity, and with cementless fixation [90]. Recovery of nerve function occurs about one third of the time. A serious vascular

injury, although rare, could be fatal.
O. Transfusions are required in 32% of hip replacements[37]. Self donation still leaves a 9% risk of transfusion of donor blood [37]. Complications of transfusions include, but aren't limited to, AIDS (1 in 2,135,000), hepatitis B (1 in 205,000) and hepatitis C (1 in 935,000). Lung injury occurs in 1/5,000 transfusions. Disease transmission can also occur with bone grafting. See Table 9 in the Appendix and refer to Chapter 3 for a more comple list of transfusion risks.
P. Component failure: failure of the materials or design of the implants is very rare. There have been recent recalls of one all-metal hip and of certain ceramic materials even though the actual failure rate of well made ceramics has been low (4/10,000, historically). An unforeseen recall may affect any medical device but rarely requires removal of the device from your body.
Q. Malposition of the components: the surgeon may fail to align the hip correctly, resulting in dislocations or excessive wear. It is uncommon to have such poor component position that revision surgery is mandated after the first dislocation.
R. Metal allergy: certain metals, usually nickel, may cause swelling and a rash. If you know that you have a severe metal allergy that causes bad rashes or blistering around jewelry or watches you should discuss this with your surgeon prior to surgery. Patients with metal allergies should avoid metal-on-metal hips until we have a better understanding of the importance of the increased blood metal levels caused by these implants.
S. Complications from any of the medicines used around the time of surgery may cause grave problems that are life threatening.
T. Blood thinners may cause bleeding into the incision or from the bowel or kidneys, especially if an ulcer or undetected cancer is present. Blood thinners may also

cause bleeding into the muscles or brain if an injury occurs or if there is an undetected vascular malformation. Bleeding, allergic reactions and injuries to organs from medications can be life threatening. Rarely, warfarin (Coumadin™) can cause gangrene and heparin can occasionally decrease platelet counts or rarely cause gangrene.

U. Bowel obstruction or perforation may be more common when using pain medications and after total joint surgery. The bowel may function poorly and develop a pseudo-obstruction, especially if the body's electrolyte balance is abnormal (Ogilvie syndrome). Bowel obstruction occurs in 1.5% of hip replacement patients, and is more common in revision surgery (2.2%) and in patients who have had previous abdominal surgery [69]. While bowel perforation is rare, it can be life threatening.

V. Hospital acquired pneumonia is more common with long hospital stays and may be difficult to treat due to bacterial resistance.

W. Urinary retention is more likely in men with prostate enlargement but can occur in women. Urinary tract infection can be caused by use of catheters used around the time of surgery.

X. Antibiotic use can lead to colon infection from overgrowth of a bacteria which can spread in hospitals.

Y. Late dislocations can occur many years after hip replacement and may relate to chronic inflammation and stretching of the hip's capsule from wear debris. This condition often requires additional surgery.

Z. Bursitis around the hip or trochanteric pain (pain on the outside of the hip) may effect up to 15% of patients after total hip replacement [89].

# 3. Rehab, results, and follow-up care after hip replacement

The hospital stay should be kept as short as possible to prevent hospital acquired infections. The maximum stay that a hospital will be reimbursed for is 4 ½ days for uncomplicated hip replacement surgery. The usual stay is 2-3 days. Physical therapy is started within 24 hours of the surgery. You will learn how to get out of bed using a leg lifter and begin to walk using a walker of crutches. Your pain medicines and blood thinner will be adjusted during that period. When your doctor feels that you are able to get around well enough to negotiate obstacles around your house, you are discharged. Home health services may be employed for the first week at home to assist with bathing, therapy, and dressing changes. Blood tests can be done by the home health nurses or at your local laboratory.

Rehabilitation hospitals are sometimes used to provide additional therapy in an environment which is relatively free from contagious diseases. CMS (Medicare) has debated the merits of this care for several years now and coverage through CMS or your private insurance is not guaranteed for admission to these facilities. Inpatient rehabilitation may afford closer supervision by physical therapists, which has been shown to result in lower dislocation rates. Current recommendations for inpatient rehabilitation facility (IRF) admission made by the government accounting office in April, 2005 favor patients who have had both hips replaced at the same time, patients over 85 years of age, and patients who are morbidly obese (body mass index over 50) [61]. Replacing both hips at one time has been associated with an increase in death rates and is not recommended by many surgeons in spite of this possible government benefit.

Nursing homes and skilled nursing facilities are often considered for patients who cannot return home if they are not accepted by an acute rehabilitation hospital. Unless a special skilled facility is dedicated to orthopedic rehabilitation these facilities are the least preferable choice at the time of discharge from the hospital. These facilities are often full of lower funtioning patients and patients with infections, often from resistant bacteria which are contagious.

- **Post-operative care after hip replacement**

   Bathing is permitted at any time if a special occlusive dressing is in place, otherwise it should be postponed for two weeks, substituting sponge bathing. Staples, if used to close the skin, are removed about ten days postoperatively. Swimming and the use of spas or hot tubs is best deferred until three weeks after surgery. You may return to work that does not require prolonged standing in three weeks and to usual standing and walking duties in six weeks in most instances.

   For a period of time after surgery you may be on the blood thinner, warfarin (Coumadin™), to prevent blood clot growth. Each week you should go to the laboratory and have a blood test (protime with INR). Either your surgeon or your primary physician will adjust this medication to keep the INR between 2-3, as recommended by the American College of Chest Physicians. Warfarin will be adjusted around each of the medicines that you usually take, but, because it has interactions with most other medications, you should call your surgeon's office or your primary doctor's office to discuss the use of warfarin (Coumadin™) with the other medications, including vitamins and herbal supplements, before taking them. Occasionally, ulcers and growths in the colon or urinary tract may lead to blood in the stool or blood in the urine. Should this occur while you are taking warfarin, stop the medicine and seek immediate medical care.

   In lieu of warfarin, injectable medications, such as Lovenox™ can be used. This medication does not always require laboratory monitoring. Blood tests are recommended if you develop small, red spots on the skin, called "petechiae."

Aspirin can also be used to prevent blood clots, but most physicians do not believe that it is as effective as other medications at preventing clot growth in veins.

If you develop swelling or pain in the calf or thigh, this can be a sign that you have developed a clot in a deep vein (deep venous thrombosis) and you should seek medical care. Should you become short of breath or have chest pain, this can represent a clot in the lung (pulmonary embolus) or a heart attack (myocardial infarction). These complications are life-threatening and you should call 911 without delay if they occur.

- **Dislocation precautions after hip replacement**

For six weeks after surgery, you will need to adhere to certain limitations to prevent dislocation of the artificial hip. The use of a special pillow between your legs, elevated toilet seats , leg lifters, grabbers, and restrictions from sleeping on the side are precautions which are believed by many surgeons to lessen the risk of dislocations. There is evidence that these stringent precautions may not affect the dislocation rate when an anterior-lateral surgical approach to the hip is used [91]. Since more surgeons use a posterior-lateral surgical approach to the hip, sleeping restrictions and various devices which prevent you from flexing the hip excessively for the first six weeks after surgery are still widely recommended. You should not flex the hip more than 90° or rotate the knee inward. This will allow the hip capsule that is repaired at surgery to heal in a tight position and prevent future dislocations.

At six weeks you need only avoid extremes of motion and you can sleep on either side. At any point after hip replacement you should not bend over the operated hip sideways to pick up objects on the side of your chair while seated and you should not work on the outside of your foot by twisting your knee inwards. These activities flex the hip while rotating it inward and are highly associated with dislocations. Don't force motion in your replaced hip. If the hip seems to reach a stopping point the neck of the prosthesis may be impinging on the cup. Forcing motion beyond this point may lever it out of place.

- **Resuming activities and long-term results after hip replacement**

You may resume driving three weeks after your surgery as long as you are not taking pain pills and you feel that your reaction time has returned. You may be asked to drive with a companion initially to ensure that your reaction time is normal. You may also choose to test yourself on a driving simulator, which may be available through some occupational therapists and at driving schools. You may be allowed to return to part-time work that doesn't require prolonged standing at three weeks. You can return to full work duties by six weeks, in most instances.

Sexual relations can be resumed whenever it is comfortable to do so. The hip should not be flexed beyond 90 degrees in the first six weeks and never more than 120 degrees, but spreading your legs is actually a very safe position for the artificial hip.

Travel is reasonably safe after joint replacement while the blood is thinned with injectable blood thinners or warfarin. After discontinuing the use of blood thinners there is a period of time that travel may place you at risk for blood clots in the veins or clots that pass to the lungs. The risk can be decreased to some extent by taking breaks to stretch your legs. Especially common with air travel, blood clots can occur with any period of prolonged sitting or bed rest. Exercise and avoiding dehydration are recommended but the use of aspirin during travel is no longer recommended by the American College of Chest Surgeons to prevent clots in leg veins during travel. Should you need to travel soon after total joint replacement surgery, you should discuss the use of injectable blood thinners with your physician.

Sports should be deferred for twelve weeks. Many

court and field sports and all contact sports are not advised. You will be able to hike, golf, play tennis, and bicycle. If you were able to ski before your surgery, you may resume skiing at this time. After a hip replacement you should not try to learn to ski or ride motorcycles due to the risk of fractures. Fractures around the artificial hip will be more difficult to fix than they would otherwise have been and dislocations of the artificial hip are more common than in a normal hip. This is addressed in the section on fractures after joint replacement.

Long-term results are better for hip replacement than for any other joint replacement surgery. Even before the use of alternative bearing surfaces, many series demonstrated average survival of the joint for over twenty years. About 98% of the patients in several studies have no pain or so little pain that they do not take medicines. About 2% of patients have a vague mild thigh pain that is usually best left untreated. About 3% of patients will dislocate an artificial hip, perhaps increasing somewhat over the years after the surgery. The book is being re-written on survivorship of hip implants with the use of new technologies and new surgical techniques. Although hip replacement has been very successful for many years now, our results should continue to improve.

- **Antibiotics for dental work and procedures**

    Antibiotics before dental and urologic procedures are recommended. The risk of an infection traveling through the blood steam after a dental procedure is about one in 10,000. Two grams of ampicillin, cephalexin or cephradine is prescribed for use 60 minutes before the procedure for patients not allergic to penicillin-type medications. For patients allergic to penicillin, clindamycin 600mg is substituted. For patients with multiple severe allergies it is probably best not to use antibiotics before these procedures unless there are risks for infection from medications that suppress the immune systems, from other diseases, or from the nature of the dental or urologic procedure.

- **Long-term follow-up with your surgeon**
Every few years many surgeons will take X-rays to ensure that the joint not loosening, not wearing excessively and not causing bone to thin around the implant. You should seek an earlier appointment if there is a new injury, swelling or loss of stability of the joint.

If you develop any change in the alignment of the foot or new pain in the groin or thigh, you should have your hip x-rayed and evaluated by your surgeon. When a hip replacement becomes loose it may cause shortening of the leg or out-toeing (walking with the foot pointing outwards). Because bone loss can be rapid with a loose implant, fixing the problem may be much simpler if it is caught early.

- **Back pain, buttock pain, and hip arthritis**
The primary goal of hip replacement is not relief from back symptoms. Back pain and buttock pain are often from degenerative spine conditions. Back pain and buttock pain may not improve with hip replacement. Hip replacement is expected to relieve groin pain, thigh pain, and sometimes the knee pain that may be associated with hip disease.

There is one small study, however, that shows that back pain may be relieved in some patients with hip arthritis who undergo hip replacement. Hip-spine syndrome is the association of back pain with the abnormal gait (limp) that is associated with hip arthritis. Partial relief from back symptoms lasting at least two years was seen in this study of patients undergoing hip replacement[143].

- **Giving the surgeon a grade**
Dr. William Harris of Harvard University developed a scoring system in 1969 that is still widely used to access results after hip replacement. The appendix contains the scoring system developed by this pioneer in hip surgery,

modified slightly to simplify scoring the range of motion of the hip. Surgeons who practice outcome-based medicine often track this score to compare preoperative and postoperative results. This helps surgeons identify factors which improve their results over time. A modified Harris hip score is included in Table 12 of the Appendix.

# 4. Technologies and techniques for hip replacement

- **Historical progress**

In the era of Sir John Charnley, who pioneered modern hip replacement in England in the early 1960's, the components were cemented into the body. This technique was durable except for the cemented polyethylene cups which tended to loosen faster than the cemented stems. Even so, many of the replacements performed by Charnley survived for over 20 years. Cementing techniques in surgery improved but cup loosening still occurred. Bone ingrowth cups produced more durability but we began to notice that bone was dissolving around the implants, especially when the wear of the plastic was accelerated.

Dissolving bone around the implants was originally felt to be a reaction to the cement and we began to replace hips without cement. The first cementless stems did not perform as well as expected. There was still the problem of dissolving bone, now called "osteolysis" instead of "cement disease." The identification of wear particles of many types as well as the biologic events that caused osteolysis lead William Harris, MD, of Harvard University, and others to focus on eliminating wear from the motion surfaces of hip replacements.

With the development of wear resistant material for use in the bearing surfaces of hip replacement, osteolysis has been greatly reduced and we believe that current alternative bearing surfaces have the potential to greatly limit this disease and make the hip replacement permanent in some instances.

- **Ingrowth femoral stems**

The most successful ingrowth stems developed have been tapered titanium stems without collars. They are as durable as cemented stems and the extreme long-term results are expected by many surgeons to outperform cemented stems. Titanium appears to allow better ingrowth than cobalt-chrome. The use of collars on femoral stems may interfere with intimate contact between the remaining stem and the surrounding bone resulting in a small incidence of stems that fail to ingrow and become loose.

- **Cemented stems**

Still in use, some polished tapered cemented stems have demonstrated durability beyond 25 years. When a polished stem does loosen and lose intimate fixation with cement, it doesn't produce large amounts of debris like the roughened cemented stems do. This simplifies revision (redo) surgery and slows down the rate of bone loss for patients who wish to postpone their revision surgery. On the basis of historical data alone, it is hard to beat a cemented stem.

- **Highly cross-linked polyethylene**

Warming the plastic while subjecting it to radiation produces chemical bonds between the long polyethylene molecules. This reduces the wear of this surface in hip replacements. At first, it was difficult to see that any wear occurs with highly cross-linked polyethylene. Refinements in our ability to measure wear have shown that small amounts of wear do occur, but these levels appear to be insufficient to trigger osteolysis (dissolving bone). The historically acknowledged threshold for wear significant enough to result in dissolving bone is 0.2mm a year, well above anything measured in highly cross-linked polyethylene cups. The wear is low enough to allow the use of large femoral heads to reduce the likelihood of dislocating the hip from impingement of the metal parts. This was not possible with the older conventional polyethylene which produced large volumes of wear with large heads. This wear-resistant plastic allows surgeons a great deal of flexibility in achieving hip stability. Large heads, oblique liners, elevated rims on liners and constrained liners have each been shown to reduce the rate of total hip dislocations in various circumstances. Polyethylene is the bearing surface for cups that gives the surgeon the most flexibility in preventing dislocations.

- **Ceramic heads and surface ceramics in hip replacement**

Even before the development of highly cross-linked polyethylene, the wear of the plastic had been reduced to below the critical threshold for osteolysis (0.2mm of wear per year) by using ceramic heads against conventional polyethylene. Metal heads show much more scratching than do ceramic heads when they are retrieved after years of use. The use of ceramics may be particularly valuable when using highly cross-linked polyethylene, which may be more sensitive to wear from scratches.

Unfortunately, solid ceramics can break and have done so at a rate of about 4 per 10,000. The risk of ceramic breakage has been eliminated with the use of the surface ceramic head, Oxinium™. The combination of Oxinium™ and highly cross-linked polyethylene appears promising but long-term follow-up studies are still required.

- **All-ceramic hips**

The lowest rate of wear in any hip articulation is from the all-ceramic hip made of alumina. The other ceramic in use for ceramic on polyethylene hip replacement is zirconia. Zirconia has not fared as well in all-ceramic joints. The fracture resistant surface ceramic, Oxinium™, is made with zirconia, and can't be used in the all-ceramic hip. Using alumina to make a socket is more difficult than other ceramic applications. It does not support elevated rims, oblique liners, or constrained liners. Refinements in our ability to measure wear have shown that small amounts of wear do occur, but the level of wear appears to be insufficient to trigger osteolysis unless a ceramic component becomes loose.

At extremes of motion, the femoral stems that the

hip ball mounts to may impinge on the ceramic, causing chipping and creating wear debris. One solution has been to protect the edge of the ceramic by shrouding it in a titanium jacket. Unfortunately, this can result in a greater tendency for impingement of the stem on the cup, further limits hip motion, and may prove to further increase dislocations of the hip post-operatively.

All-ceramic hips can squeak and make other noises. Rarely, they fracture. The rare fracture of solid ceramics results in so many small fragments that they can't all be discovered and removed at the time of revision surgery. Results from revisions for broken ceramic components are among the worst in total joint surgery [92].

Why shouldn't everyone have an all-ceramic hip? The very elderly should not take the chance. The extraordinary durability of this implant isn't required in this instance and with increased risks for falls, a ceramic fracture would be a poor addition to their golden years. In obese patients we would anticipate more trouble with obtaining perfect cup position due to the difficult exposure and lack of surgical landmarks. An imperfectly placed cup is in an invitation to disaster in the forms of chipping, impingement and dislocation. Patients who do not give up activities which place them at risk for trauma and falls should be aware of the possibility of ceramic breakage. This includes motorcycle riders, ATV riders, bicyclists, aggressive skiers and alcoholics.

Who is best suited for an all-ceramic hip? Golfers, hikers, swimmers, and young people who are not engaged in activities that will lead to high speed collisions and falls will be best served by an all-ceramic hip. It could last them forever and that is the goal with this articulation. The rest of the young, active patients may be better served with metal or Oxinium™ on highly cross-linked polyethylene or with all-metal hips.

- **All-metal hips**

The second lowest rates of wear are seen in all-metal articulations. The only reason that this articulation is not in wider use is the concerns about metal levels in the blood streams of patients who underwent surgery with these implants. All-metal hips have been associated with abnormal immune cell proliferation around blood vessels in the tissues around the hip that are not seen in metal on polyethylene hips, findings that show an allergic response (type IV, delayed hypersensitivity) [93, 94, 95, 96]. Metal-on-metal hips have yet to be shown to improve the durability of hip replacements [95, 96]. The disease, osteolysis, that metal-on metal bearings are supposed to prevent still occurs in patients that have receive these types of implants [95,96]. There is also evidence that all-metal hips are associated with metal allergies [97]. Gene mutations have been seen more frequently in the blood (lymphocytes) after all-metal hip replacement [98] but this has not been shown to be associated with cancers in patients with metal-on-metal hips [99]. Most surgeons are exercising caution with all-metal hips due to a concern about unforeseen long-term biological consequences of walking around with increased levels of cobalt, chrome or molybdenum in their blood. Staying "one trend behind" may prevent unforeseen problems with the use of all-metal hips in large groups of patients because rare problems are statistically more easily identified in large groups.

These implants have been used in small numbers since 1938, even before we used polyethylene. The only recall of an all-metal cup was unrelated to the metal. Due to a production error, small amounts of oil were left on the cups from the manufacturer Sulzer™, blocking bone ingrowth in some patients, and requiring revision surgery in over 1700 of the 17,000 cups placed.

More detailed study of all metal hips, to include resurfacing hip replacements, is required to routinely recommend them and long-term studies will be required to insure that issues relating to metal hypersensitivity and dissolving bone are better understood.

- **Resurfacing arthroplasty**

Resurfacing hip replacement involves placing a metal cap on the ball of the femur and implanting a metal socket. It was proposed to conserve bone around the neck of the femur, most of which is removed during conventional hip replacement. If the socket is not replaced, the cartilage around the socket wears out and shortens the lifespan of this implant. Results from hemiarthroplasty (replacing the ball but not the socket) have been inferior to total hip replacement (replacing the ball and the socket). However, replacing the socket in a resurfacing replacement makes the surgery little more "bone-conserving" than conventional total hip replacement in the view of many surgeons.

Resurfacing hip arthroplasty uses metal-on-metal for its bearing surface. All metal hips have been associated with abnormal immune cell proliferation around blood vessels in the tissues around the hip that are not seen in metal on polyethylene hips [93, 94, 95, 96]. These findings are consistant with delayed hypersensitivity (allergy) to metals. Patients can develop allergies to metals in metal-on metal hips [97]. Also, gene mutations have been seen more frequently in the blood (lymphocytes) after all-metal hip replacement [98]. This finding had previously only been reported in failed metal on polyethylene hips, which also created metal debris [93]. Most surgeons are exercising caution with all-metal hips due to a concern about unforeseen long-term biological consequences of walking around with increased levels of cobalt, chrome or molybdenum in their blood.

Resurfacing implants can wear, become loose, and cause osteolysis (to a lesser degree than hips with plastic articulations). They also can cause hip fractures that occur below the level of the resurfaced ball of the hip joint if that portion of the bone is violated by the surgery, progression of the avascular necrosis (loss of blood supply top the bone) beneath the metal implant, impingement of the bone on the cup, or if dissolving bone (osteolysis) occurs after metal-on metal hip replacement [95, 96]. Because these implants have had a shorter life span than total hip replacements when cement is used [100] and require revision of the socket to revise them to conventional total hip replacement, many surgeons are not yet recommending this type of hip replacement until long-term results with cementless resurfacing are available and metal allergy concerns are sorted out.

- **Avascular necrosis (AVN)**

    Also, called osteonecrosis, this process leads to destruction of the bone from loss of its blood supply in the femoral head (ball of the ball and socket). Most commonly, no cause of the loss of blood supply can be discovered. Other common causes of this disease are the use of steroids or chemotherapy, trauma, and alcohol abuse. AVN can occur in childhood, and is called Legg-Calve-Perthes' disease. In its early stages, AVN has been treated with decompression through drill holes and with bone grafting, to include vascular grafts from the patient's ankle. Success has been variable with these techniques. A recent study showed that the use of medications for osteoperosis (bisphosphonates) may prevent the collapse of the femoral head that occurs when the blood supply returns and the dead bone is resorbed by osteoclasts, the cells in the body that remove old bone [5]. A second study also showed a lower rate of collapse of the femoral head in patients taking aldendronate (Fosemax™) [101]. Ironically, the use of bisphosphonates has been associated with osteonecrosis of the jaw, leading to serious infections following dental procedures. This rare disease has been seen mostly in bone cancer patients taking these drugs in intravenous form and only isolated reports of patients using oral bisphosphonates for osteoporosis have been published [102].

    It has been suggested that AVN should be treated

differently from arthritis because it occurs in younger people and the results of hip replacement have not always been as good in AVN as they were in osteoarthritis. Replacement of only the femoral head (hemiarthroplasty or hemiresurfacing) fails more often than total hip replacement [103]. Resurfacing arthroplasty, discussed earlier in this chapter, has not been shown to be as durable as total hip replacement to date. Many surgeons avoid the metal-on-metal designs used in resurfacing arthroplasty due to concerns about biological effects of systemic metal levels and metal allergies. Long-term results with current designs are required to understand the exact role of resurfacing designs in the treatment of AVN and osteoarthritis. For these reasons, many surgeons continue to recommend total hip replacement for their patients with AVN.

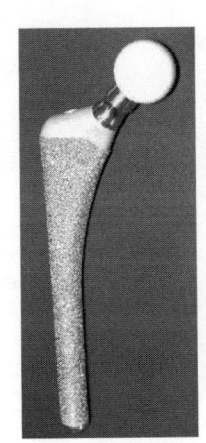

- **The flexible composite hip**

Another development in hip implants is the Epoch™, a composite stem. Released by the FDA in 2002, this stem has potential to prevent two problems in hip replacement. First, loss of bone adjacent to large, stiff implants occurs due to shielding of the bone around the metal implant. Because the stiffness of the implant increases exponentially with the size of the implant, this is especially true for patients with large diameter bones. The Epoch™ is closer in its flexibility to the bone around it and appears to lessen this undesirable effect. Secondly, when a stiff implant is placed in the bone, it may cause thigh pain after hip replacement. This new stem has been associated with a low incidence of thigh pain. There is no long-term data available for this implant yet.

This implant has a unique appearance on x-ray because there are three layers to it. The inner layer is cobalt-chrome. The middle layer is a material called poly-aryl ether ketone (PAEK). The outer layer is now made of tantilum which supports bone ingrowth.

- **Tantalum, another metal for bone ingrowth**

A different elemental metal, tantalum, is used in Trabecular Metal™ implants. This new material may allow even better bone ingrowth than titanium. There is also some potential for soft tissues to grow into this material, making it useful for situations where bone has to be replaced with a metal implant. Trabecular Metal™ has been used primarily in revision (redo) surgery but increasingly it is finding its way into mainstream hip replacement surgery. It also has potential for use in implants used in tumor surgery where more bone must be removed because this material may allow muscles and tendons to ingrow into it. Once again, we will await the results from long-term studies of these implants.

- **Mini-incision surgery**

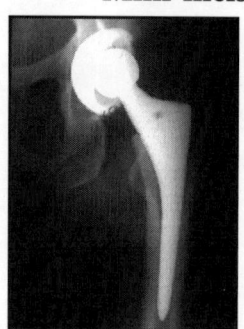

Many surgeons began to use progressively smaller incisions a couple of years before there was any discussion of mini-incision surgery or minimally invasive surgery. We began to notice that the occasional patient had a cup that was too vertical on their postoperative x-rays. A vertically positioned cup may lead to more wear or to dislocations, requiring revision surgery.

When many surgeons didn't notice any improvement in the consistency of their cup position as they gained more experience with mini-incision surgery, they became concerned that the surgery would require computer navigation which would increase costs and lengthen the surgery.

Another concern is that the math of the incision does not add up. A cup with a two inch diameter would have a circumference of 6.28 inches ($\pi$ times the diameter). A 2½ inch incision would have a perimeter of only 5 inches. There is a risk for skin injury with this surgery as well as hidden injuries to deep muscles or to the hip's capsule.

A group of university surgeons studied mini-incision hip replacement and found results that were contradictory to those reported by the surgeons who popularized mini-incision surgery [75]. These surgeons, who were all well trained and experienced, found higher rates of skin healing problems and poorly positioned cups as well as ill-fitting stems. Surprisingly, the mini-incision patients had no less blood loss, no difference in postoperative rehabilitation and no shorter hospital stays. Because they selected slender patients for the mini-incision surgery who were healthier than the group of standard incision patients, the results were particularly disappointing. In another study, a surgeon who had previously performed over 300 mini-incision surgeries prior to starting the study randomized his patients into standard incision versus mini-incision groups and found no change in blood transfusion rates, no easier rehabilitation, no better results, and no shorter hospital stays. Importantly, this surgeon did not have more complications in his mini-incision group than he had in his standard incision group[104]. We can infer from this study that, even in the hands of a surgeon good enough to avoid complications through small incisions, we can't prove a benefit from small incision surgery other than a smaller scar. A smaller scar does not equate to a more cosmetically satisfactory incision. A study comparing the cosmetic results of mini-incision scars to standard incision scars rate 30% of the mini-incision scars as poor but only 7% of the standard incision scars as poor [105].

Minimally invasive two incision surgery has been associated with higher complication rates (14% versus 5%) and more reoperations (5% versus 1%) in one study [106]. However, the pioneer of this surgery, Dr. Richard Berger, has reported excellent results with two incision surgery with no change in complication rates compared to standard incision surgery but consistently reduced hospital stays with 85%

of the patients going home on the day of surgery [107]. The surgery is demanding and specialized training for the surgeon has been recommended. Concern over muscle injury with this technique exists [108] and this type of surgery cannot be safely done on obese patients.

There is experience that supports the use of these techniques, but it is clear that this is not yet a viable option for all surgeons. One half of all hip replacements in the United States are done by surgeons who do ten or fewer hip replacement surgeries a year [109]. This may not be enough volume to become confident in mini-incision or minimally invasive surgery and the learning curve may compromise some of the results of this surgery if the surgeon is new to it. When the data supports this modification, our surgeons will be trained in minimally invasive techniques in the universities before they begin practice in private hospitals. In the interim, be assured that the jury is still out on this subject and that hip replacement done with conventional incisions is the most successful surgery in all of medicine at improving a patient's quality of life. Additional scientific work is required to routinely recommend these procedures although each has promise as a technology in development.

The ability to perform hip replacement through very small incisions without a higher complication rate may best be regarded as a marker for a high level of surgical skill. Even in the best hands, the incision must be extended in some instances due to intraoperative findings and the need for better visualization. Obese and very muscular patients are at the greatest risk for complications with small incision surgery.

- **Surgical approaches to the hip**

The anterior approach to the hip is a safe approach in the hands of a surgeon experienced in it. A specialized operating table, not available in most hospitals, has been recommended for this surgical approach. The dislocation rate with this approach is 0.61% in the hands of one experienced surgeon [110]. The posterior approach with repair of the capsule and tendons is associated with a 0.85% dislocation rate in the hands of one experienced surgeon who used the more conventional posterior approach [85]. Popular among surgeons who specialize in pelvic surgery, the anterior approach can reduce dislocations but may not be a good choice for orthopedic surgeons without such specialized training.

Most surgeons use either a posterior-lateral approach or an anterior-lateral approach. Historically, most surgeons have believed that the anterior-lateral approach has been associated with fewer dislocations but could result in a limp if the muscles that were detached fail to heal after they are repaired. The posterior-lateral approach combined with repair of the hip's capsule has an acceptably low dislocation rate and is felt to be less invasive by many surgeons because no powerful muscles are detached. Revision surgeons use both of these as well as the direct lateral approach and a procedure that opens up the top of the canal of the femur, the extended trochanteric osteotomy. Failure to unite an osteotomy can require more surgery but the use of an osteotomy may be required for bone deformity and the removal of well fixed stems.

Asking surgeons to change their surgical approach is like asking them to change their religion. Most surgeons are much more comfortable with one approach and asking them to modify this is an invitation to disaster. Fortunately, however, there is no overwhelming benefit of one surgical approach over another.

- **High volume surgeons and hospitals**

There is data that complication rates are lower among surgeons and hospitals that perform higher volumes of total joint surgery. For instance, the dislocation rate is 4.2% in low volume surgeons' hands compared to 1.5% in high volume surgeons' hands [109]. The additional margin of safety is not overwhelming and you can find orthopedic surgeons in most areas of the country that will have outstanding track records with most joint replacement procedures.

# Chapter VI. Medical Issues after Joint Replacement

## 1. Anemia

Most patients have anemia (low red blood cell counts) after joint replacement. This can lead to a loss of energy. If symptoms persist, a blood count should be obtained to ensure that they are from the anemia and not side effects of pain medications or symptoms of depression which can also lead to unusual fatigue. Patients with cardiac and pulmonary disease may need transfusions with cell counts that are closer to normal than the cell counts that would lead to transfusion in healthy patients. Dizziness can occur with sudden blood loss, when heart rate is high or low, and with dehydration. Seek medical care immediately if you develop new dizziness. Even those opposed to a blood transfusion should undertake an evaluation for dizziness to rule out cardiac or neurological causes and to obtain treatment for dehydration if necessary.

Mild or moderate anemia is treated with iron supplements, especially in women. After a lifetime of having menstrual periods, women have low iron stores in their bone marrow. This is not the case for men and men may not require iron supplements after surgery. As iron may cause hard stools, you may also wish to use a stool softener in addition.

## 2. Blood clots (deep venous thrombosis, DVT)

Blood clots frequently form in the deep leg veins and pelvis veins in total joint patients. There is often a vague new pain in the calf and more swelling than is expected from the surgery. When there is any doubt about blood clot formation an ultrasound test called a "duplex DVT study" should be ordered by your physician. It may be ordered for both legs as 10-15% of clots occur in the unoperated leg [111]. A duplex DVT study will not usually find clots in the pelvis which may form after hip replacement.  One study showed that 10.4% of patients undergoing hip replacement had pelvic clots [87]. To discover these clots either a CT scan or MRI scan would be required (CT with contrast or MR venography).

Blood thinners are often continued after discharge from the hospital.  42% of the clots (DVT) after knee replacement and 76% of the clots after hip replacement are discovered after the patient is discharged to home [55]. The average time to discovery of a blood clot (DVT) is 7 days after knee replacement and 17 days after hip replacement [55].

Blood clots (DVT) are treated with blood thinners, usually for 3-6 months.  If you have a personal or a family history of frequent blood clots, additional investigation may be performed to determine if you have a hereditary or an acquired cause of excessive clotting.  The diseases associated with excessive blood clot formation (called "hypercoagulability") may require lifetime use of a blood thinner, typically warfarin (Coumadin™), to prevent potentially fatal blood clots to the lungs, called "pulmonary emboli." The testing for excessive blood clot formation is in the appendix as is a technical section on prevention of DVT.

Clotting problems can be hereditary.  Since the screening tests for excessive clotting (throbophilia-hypofibrinolysis) are expensive, the most cost-effective way of preventing clots to the legs and lungs (DVT and PE) is to treat all total joint patients with anti-coagulants.  A study from the Hospital for Special Surgery in New York looked at the patients undergoing total joint surgery who had DVT or pulmonary emboli in spite of treatment to prevent clots to investigate how many had abnormal clotting.  Over 50% of these patients had heritable clotting abnormalities.  The three labs and the percentages of patients with leg clots or pulmonary emboli with abnormalities discovered at least three months after tratment for their clots in this study were antithrombin III (12%), protein C (21%) and prothrombin gene mutation (23%) [112].

## 3. Blood pressure changes

Spinal or epidural anesthesia can decrease blood pressure after surgery. Occasionally, the blood pressure will remain a bit lower for several weeks after surgery and will require adjustment of blood pressure medications over that period. Unfortunately for our patients with hypertension (high blood pressure), this change is only temporary and eventually the blood pressure medications will have to be resumed. More study is required to understand this phenomenon.

Anemia does not cause low blood pressure unless the blood volume is decreased. If it has been several days since surgery the blood volume is normal unless some dehydration is also present. Drink plenty of fluids unless you have been told not to because of electrolyte problems such as low sodium level on blood tests.

## 4. Cardiac problems

Chest pain can be caused by a variety of medical problems but two of them, heart attacks and blood clots to the lungs (pulmonary embolus) are emergencies and should precipitate a call to 911.

Because of fluid shifts after surgery the upper chambers of the heart (the atria) may be stretched and cause an irregular heart rhythm, called atrial fibrillation. This heart rhythm has been associated with strokes and heart attacks and can lead to high enough heart rates that the blood pressure drops.

## 5. Anorexia (loss of appetite)

Usually a side effect of narcotic pain medications, you may lose interest in food after surgery. Substituting non-narcotic pain pills will generally solve the problem but, in extreme instances, medications that stimulate the appetite may be recommended to prevent malnutrition.

Malnourished patients have more problems with wound healing and infections.

## 6. Constipation

This can be another side effect of narcotic pain pills. Constipation is common after joint replacement. Avoiding narcotics is prudent if constipation is troublesome. Unless you have been advised to limit water intake, drink water to avoid dehydration. Also, try to stay mobile as inactivity contributes to constipation. You may safely use fiber supplements, even if you are taking the blood thinner, warfarin. Medications to move the bowels, such as milk of magnesia, may be used for short periods of time unless you have kidney failure. Long-term use of bowel stimulants is not advised.

When constipation is severe, bowel obstruction may occur. You should call your physician if you have not had a bowel movement for 3 days. Seek immediate medical attention if your constipation is associated with vomiting, fever, or abdominal pain.

## 7. Diarrhea

Hospitals may transmit bacteria that cause colon infection to their patients. The bacteria, *Clostridium Difficile*, can be contagious. This illness, colitis, is more common in hospitalized patients who are given antibiotics. If you have diarrhea you should contact your physician who can order a test for this disease. *C. Difficile* colon infection can be difficult to eradicate and may require the use of antibiotics specific to this disease. The use of inappropriate antibiotics can make the disease worse.

## 8. Breathing problems

4% of total joint patients have undiagnosed sleep apnea before they undergo total joint surgery and an additional 2.7% are known to have sleep apnea before joint replacement surgery. This means that 6.7% of patients undergoing joint replacement have this disorder which can make the use of pain medications dangerous [113]. This malady, more common in overweight patients, causes loss of breathing for periods of time during sleep and leads to feeling tired even after a full night's sleep. Heavy snoring is another symptom of sleep apnea. Unfortunately, the respiratory drive may be further depressed by the narcotic pain and sleeping medications given after the surgery. Respiratory depression is a cause of death after major surgery. Sleep apnea is often discovered after surgery when patients wear an oxygen saturation monitor. Supplemental oxygen or the use of an apparatus to assist with breathing (CPAP or BiPAP) may then be prescribed.

Patients with known moderate or severe sleep apnea are at risk of death with joint replacement surgery and special precautions are required to minimize this risk. These include avoiding strong narcotic and sleeping medications as much as possible, use of CPAP or BiPAP, continuous oxygen saturation monitoring and elevating the head of the bed [114]. Since so much sleep apnea is undiagnosed before surgery, many surgeons order oxygen saturation monitoring for all of their patients.

At the time of surgery, fat from the bone marrow travels to the lungs and can decrease oxygen exchange. This often leads to lower blood oxygen levels, especially in patients with heart failure, lung disease or sleep apnea. Again, oxygen is prescribed. The treatment of underlying heart or lung disease and the passage of time should allow you to discontinue your oxygen under the supervision of your medical doctor.

The use of supplemental oxygen around the time of surgery has been shown to be associated with a lower infection rates after abdominal surgeries. It is not yet clear if this will hold for joint replacement surgeries also. It isn't terribly inconvenient to use oxygen in the hospital, but using oxygen at home may require the attention of your primary care physician or a pulmonary medicine specialist. These specialists periodically test your oxygen levels and certify to Medicare or your private insurance company that you still require oxygen. The question, "How long do I need to use oxygen?" should be discussed with these physicians. Your orthopedic surgeon typically does not have equipment to test lung functions and oxygen levels, nor do most orthopedic surgeons have expertise in pulmonary medicine.

Pneumonias can be acquired in hospitals, especially in fragile patients with long hospital stays. Fever with a cough, shortness of breath, and wheezing require urgent evaluation by your primary care provider or an emergency physician. Shortness of breath is a medical emergency. It can be caused by blood clots in the lungs (pulmonary emboli), heart failure, asthma, or pneumonia, each of which can be rapidly fatal.

## 9. Shortness of breath and chest pain

These symptoms demand emergency medical treatment. Call 911. Blood clots to the lungs (pulmonary emboli) and heart attacks are more common after joint replacement and can be fatal. While there are many other causes of these symptoms, it is best to assume the worst and call for an ambulance should you have chest pain or shortness of breath.

## 10. Swelling

It is rare not to have some degree of swelling in the operated leg. Lymphedema, a typically harmless collection of fluid, is the usual cause. Lymphedema will improve with elevation and the use of compression stockings. Swelling that is not intermittent may reflect a clot in the deep veins of the leg. These clots often cannot be appreciated by examining the legs and require testing to diagnose. A quick painless ultrasound test will detect many clots, but clots in the pelvis, which may occur after hip replacement, require a CT scan or MRI (MR venography) to diagnose.

Swelling in both legs is more common in elderly patients and may require the use of a diuretic medication (water pill). This can also occur in the setting of blood clots, kidney disease, malnutrition, and heart failure. Only your physician can sort this out.

## 11. Bladder symptoms

Bladder catheters are often required during and after surgery and can cause urinary tract infections. Symptoms include frequent urination, burning and odor. Left untreated this infection can involve the kidneys and spread to the bloodstream, causing life-threatening sepsis (urosepsis). Your urine should be tested and you should be placed on an antibiotic for treatment of a urinary tract infection if you are found to have one. Simply drinking cranberry juice is not sufficient to treat a disease that can spread to the artificial joint through the blood.

More frequent urination is common after major surgery as your body gets rid of excess fluids given around the time of surgery. People with diabetes should test their blood sugars more frequently after surgery to ensure that the cause of urinary frequency is not high sugar levels.

Middle aged men often experience symptoms from enlargement of their prostates. After surgery they may have difficulty voiding. Pain medicines may magnify symptoms of prostatism after surgery. When caused by prostate enlargement, urinary retention is often treatable with oral medications. Severe urinary retention requires temporary replacement of the bladder catheter to prevent damage to the bladder. Antibiotics may be prescribed to prevent bladder infection if the catheter is replaced because of urinary retention but antibiotic prophylaxis is not essential.

Urinary retention with leg weakness requires further

evaluation to insure that the nerves to the bladder are not compressed after spinal or epidural anesthesia, especially when blood thinners are used. These rare problems (epidural hematoma and epidural abscess) are caused by blood clots or infected tissues pressing on the nerves to the bladder and legs.

Blood in the urine may be seen when you are on blood thinners. Blood in the urine should always be evaluated by your physician. It can signify a tumor in the urinary tract. Small amounts of bleeding shortly after insertion or removal of a catheter can safely be ignored.

## 12. Ulcers and GI bleeding

Ulcers are more common after physical stresses, such as surgery. Vomiting of blood or material that looks like coffee grounds and black tarry stools are signs of a bleeding ulcer and require immediate evaluation in the emergency room. Bright red blood in the stool can be from the colon or from hemorrhoids and also requires urgent evaluation. Finding small amounts of blood on the toilet tissue is less cause for concern. Often caused by hemorrhoids, treatment is with stool softeners unless symptoms continue.

Sometimes stomach pain and feeling full after small amounts of food consumption may occur from ulcers before bleeding occurs. These symptoms warrant further evaluation by the primary care physician or gastroenterologist. Other times life-threatening bleeding occurs without warning.

## 13. Confusion and neurological disorders

The use of anesthetic medications, pain medications, and sleeping pills can cause confusion, disorientation, hallucinations, and lack of concentration or coordination. The fat that is liberated from the bone marrow during surgery is primarily deposited in the lungs. The fat that isn't deposited in the lungs can travel to the brain and cause neurological symptoms. The fat that is deposited in the brain from major bone injury or surgery can cause subtle changes for longer periods of time. The same is true for open heart surgery. The natural history of these changes deserves more study, but these changes are rarely a source of complaint for total joint surgery patients.

Because strokes can be precipitated by heart rhythm changes after surgery and bleeding in the brain can occur with the use of blood thinners, any neurological symptoms require evaluation. Fortunately, however, most of these problems turn out to be from the narcotic pain medications or sleeping pills which can be stopped or decreased with quick resolution of symptoms.

Infections can cause mental status changes. Systemic infections can occur with pneumonia, urinary tract infection (urosepsis) and wound infections. Treatment of these infections is urgent and requires further evaluation by your physician.

## 14. Depression

The frequent occurrence of depression after joint replacement may relate to the use of pain medications or to other physical and psychological factors. Avoiding narcotics is prudent but may not be possible soon after surgery. Depression can result in loss of appetite, poor sleep, memory problems, and fatigue. It may mimic many other conditions that occur after total joint surgery. Depression may be associated with anxiety or panic attacks which mimic serious medical problems such as pulmonary embolism.

Depression after joint replacement surgery is often associated with a change in the chemicals called neurotransmitters in the brain. Changes in the levels of these chemicals can be caused by pain medications. Occasionally, your physician may recommend the use of anti-depressant medications that increase the amounts of the chemicals norepinephrine and serotonin, which cause depression when their levels are low. Once you have started this type of medication, it should not be abruptly stopped without medical supervision, as rebound depression may occur. If you have thoughts of suicide, seek medical attention without delay.

## 15. Insomnia

Inability to fall asleep or to stay asleep is frequent after joint replacement surgery. This can result from anxiety, sleep apnea, or inappropriate use of medications. Two types of medications can contribute to insomnia. Sleeping pills and narcotics (pain pills) cannot be taken for more than one week at bedtime to induce sleep without causing rebound insomnia. Sleep may be poor after total joint surgery. Continued use of pain or sleeping medications will not improve sleep over the long run. It is safer to use over-the-counter medications such as Benadryl™-based sleeping pills or low doses of prescription anti-depressant medications to obtain sleep than it is to use your pain pill to get to sleep. Discuss these options with your physician before treating yourself with pain pills at bedtime to get sleep.

Lunesta™ (eszopiclone) is a sleeping pill that has been FDA approved for longer term use. Results with postoperative patients with insomnia are not yet available. Benedryl™ (diphenhydramine) or Elavil™ (amitryptiline) may also be used short-term as sleep aids. These medications are useful as bridging therapy but may result in increased falls in the elderly and are best avoided for long-term use.

Observing good sleep hygiene is the best long-term solution. Avoid caffeine and alcohol within 4 hours of bedtime. Obtain exercise in the daytime and keep bedtime consistent, using sunlight as your biological clock when you can. Avoid naps and sleep only in your bed. If you are unable to sleep after 30 minutes in bed, get up and engage in a mundane activity until you feel tired.

# Chapter VII. FAQs (Frequently Asked Questions) about Joint Replacement

**How soon before I get up after surgery?**
You will get out of bed soon after muscle function returns from anesthesia. Longer periods of bed rest can cause fever from atelectasis (incomplete expansion of the lungs), blood clot formation, bedsores, elevations of blood calcium (which can cause confusion) and deconditioning (weakness and light-headedness when you do get out of bed).

"…25,000 out of every million of people we have must die every year. That amounts to one-fortieth of our total population. Out of this million ten or twelve thousand are stabbed, shot, drowned, hanged, poisoned, or meet a similarly violent death in some other popular way, such as perishing by kerosene lamp and hoop-skirt conflagrations, getting buried in coal mines, falling off housetops, breaking through church or lecture-room floors, taking patent medicines, or committing suicide in other forms. The Erie railroad kills from 23 to 46; the other 8445 railroads kill an average of one-third man each; and the rest of that million, amounting in the aggregate to the appalling figure of nine hundred and eighty-seven thousand six hundred and thirty-one corpses, die naturally in their beds! You will excuse me from taking any more chances on those beds."

Mark Twain, "The Danger of Lying in Bed," 1906, Harper and Brothers, New York and London

### How long will I be in the hospital?

The hospital stay is kept as short as possible to prevent complications from contagious diseases in the hospital. Hospital-acquired infections such as pneumonia, diarrhea, and wound infections are not rare and often involve resistant bacteria. Shortening the hospital stay appears to lessen the risks of these complications.

There are several options on leaving the hospital. You can return home with the assistance of home health services for therapy, dressing changes, lab testing, and assistance with bathing if required. Intensive outpatient therapy is a good solution for the healthiest patients.

Physicians are often pressured by insurance companies to have the rehabilitation performed at a nursing home or skilled nursing facility to reduce their costs. This may not be appropriate as some of these facilities have significant problems with contagious diseases and resistant bacterial infections - the very problems that your short hospital stay is supposed to prevent. Some subacute facilities serve primarily very elderly, debilitated patients and those with profound dementias (Alzheimer's disease, etc.) and are often not appropriate for total joint patients.

Sometimes, you can be admitted to an acute rehab hospital for 24 hour care. There you will be supervised by a doctor who specializes in physical medicine & rehabilitation for a stay of several days at which time you will be independent in your activities of daily living, such as climbing stairs, bathing, and dressing. Your surgeon is no longer primarily involved in your care at these facilities. Your care is turned over to a specialist in rehabilitation who will consult your surgeon only if circumstances warrant it. Rehabilitation hospitals are sometimes used to provide around-the-clock care and additional therapy in an environment which is relatively free from contagious diseases. CMS (Medicare) has debated the merits of this care for several years now, and admission to these facilities is not always covered by CMS (Medicare) or private insurance companies. Current recommendations for inpatient rehabilitation facility (IRF) admission made by the government accounting office in April, 2005, favor patients who have had both knees or hips replaced at the same time, patients over 85 years of age, and patients who are morbidly obese (body mass index over 50) [61].

**When can I resume my normal activities?**

With knee replacement you may bear full weight on the leg immediately. Many hip replacements are cementless and it may be recommended that you limit weight bearing following surgery for a period of time. In the case of 50% weight bearing, you can estimate what this feels like by placing the operated leg on a scale and pressing down until the scale reads half of your weight. Protected weight bearing is believed to encourage bone growth but may not be required if your surgeon is able to obtain excellent stability of the implants. Your surgeon will let you know how much weight you can put on the operated leg. It is best to listen to your body: if it hurts to bear full weight, stay on the walker or crutches until it becomes painless.

There is a chance of dislocating an artificial hip. Historical precautions against dislocations include the use of elevated toilet seats, grabbers, and sleeping with a pillow between the legs, and avoiding sleeping on your side, sitting in low chairs without arms on them, and flexing the hip more than 90 degrees. However, there is variability from surgeon to surgeon regarding which of these precautions a patient should follow. Avoiding low chairs, flexion of the hip more than 90 degrees, and crossing the legs may allow the hip's capsule to heal more completely over the 6 weeks after surgery.

With knee replacement you should not sleep or rest with a pillow behind the knee for 12 weeks, as a permanent contracture of the knee may develop and prevent full extension. This can also occur if you sleep on your side with the knee bent.

You may not be allowed to drive for three weeks after surgery. Even then, you should not drive if you are taking narcotics, having periods of confusion or forgetfulness, feel that the leg is weak, or if you are unable to flex the knee at least 90 degrees and extend it fully against gravity. If you are in doubt, ask your surgeon or therapist.

Most patients will not be able to return to work more than 4 hours a day until 3 weeks, full-time at 6 weeks. You should not resume walking for exercise for 6 weeks. Gentle water aerobics and heated pools or hot tubs are permissible at three weeks if the incision is fully healed. Golfers can work on their short game at three weeks, go to the driving range at four weeks and resume play at six weeks. Tennis, weight training, bowling, and skiing are best put off for 8-12 weeks.

**How do I take my blood thinner (warfarin)?**

This is taken each evening (no more than once a day), as directed by your physician. Everyone's dose is different and you need to have a blood test done weekly for the first weeks after returning home to ensure that your blood is not too thick or too thin. The lab will report these results to your physician so that your dose can be adjusted. You should not start any new medicines while taking warfarin (Coumadin™) without discussing it with your physician. Warfarin interacts with most medicines including herbal medicines and vitamins and this can result in blood clots if the blood is too thick or in life-threatening bleeding if the blood becomes too thin. Because of bleeding risks you should avoid falls and injury while on this medicine and seek medical attention should an injury occur. You will be on Coumadin™ for 2-6 weeks in most instances. Should you develop a blood clot in a vein, you will be left on warfarin for 3-6 months. Those with blood clots that traveled to their lungs and conditions which cause excessive clot formation may be left on a blood thinner permanently.

Injectable blood thinners are more expensive and

require that you learn how to inject yourself. These are used for variable periods of time after surgery, and you may be safe without blood testing for the period of time that you will be using these products (Lovenox™, Fragmin™, Arixtra™, and Innohep™).

### Will I set off metal detectors?

Yes, now that detector sensitivity has been increased following the need for increased airport security, total joint implants will set off metal detectors. Should it happen to you, they will be likely to ask you to empty your pockets and manually scan the area of your replaced joint. They probably will not honor cards or letters that show you have had a joint replacement, as cards would be nearly as available to a terrorist as they would be to you. In the United States, where 500,000 joint replacements are done annually, airport security is familiar with your situation.

### Do I need to take antibiotics for dental work?

The risks of infection in a total joint from a dental procedure are about 1/10,000. Two grams of ampicillin, cephalexin or cephradine is prescribed for use 60 minutes before the procedure for patients not allergic to penicillin-type medications. For patients allergic to penicillin, clindamycin 600mg is substituted. For patients with multiple severe allergies it is probably best not to use antibiotics before these procedures unless there are risks for infection from medications that suppress the immune systems, from other diseases, or from the nature of the dental or urologic procedure.

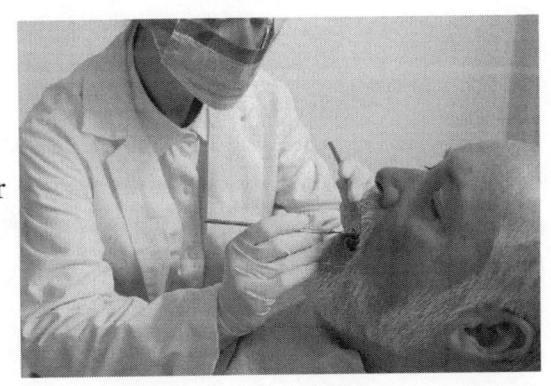

**Might I be having an allergic reaction to the metals in my joint replacement?**

This is very unlikely but not impossible. Many surgeons do not believe that they have ever seen a patient whose results were compromised from metal allergies, although all metal articulations may be associated with a higher rate of metal allergies [97]. If you have a rash and swelling over the artificial joint and you have a history of metal allergy, it may be reasonable to perform tests to investigate metal allergies. Skin patch testing for metals can be done by a dermatologist or allergist but are difficult to interpret. There is now a program for blood testing (lymphocyte proliferation testing) through Rush Medical Center in Chicago (Ms. Zheng Song 313-576-4511, CPT code 86353) to evaluate metal allergies further. Results from blood testing for allergies are also difficult to interpret and additional investigation is expected to shed more light on this complex subject. Most patients test positive for tests for allergies to bone cement (PMMA, poly methyl methacrylate). This finding is of doubtful significance.

**Are implants used after they have been recalled?**

Once a recall is made, these implants are physically removed from our hospitals. Most recalled implants actually do not cause problems that require any additional surgery. This is the case with a recent voluntary recall of certain ceramic hips in the US following an increase in the fracture rate of these devices in Europe. Fracture rates in the recalled lots were 3-5%. Historically, the risk of ceramic fracture has been only 4/10,000 and solid ceramics are again safe. The cementless versions of surface ceramic implants (oxidized zirconium) failed to ingrow in some instances and were voluntarily recalled by their manufacturer. As many surgeons do not use cementless total knee implants this recall did not affect most patients.

About 10% of the Sulzer™ metal-on-metal hips

have had to be revised due to failure of ingrowth into the acetabular shells as a small amount of oil was left on them in the manufacturing process. These implants are again available and appear to be safe after the error in manufacturing was recognized and corrected.

One batch of DePuy™ polyethylene (plastic components) produced for use in hip and knee replacement components were sterilized with irradiation (x-rays) in the presence of air, rather than an inert gas. This caused the polyethylene surface to oxidize and lead to a small number of premature failures.

Any device may be recalled if a safety issue is identified. It is best to be conservative in the application of any new technology. One good strategy to avoid recalls is to not use new technologies needlessly. An elderly patient can expect excellent durability from the use of time proven conventional technologies and surgical techniques. Your surgeon has the expertise to make the appropriate choices with you.

**My total joint still has pain. Has something gone wrong?**

There is pain in about 5% of total knees and in 2% of total hips in most series. It is very rare to have pain that is as bad as the preoperative pain three months after surgery and, should this be the case, testing for infection is undertaken. A careful exam and X-rays are obtained. A blood test is ordered. If the C-reactive protein (CRP) and erythrocyte sedimentation rates are normal, infection is unlikely (97% sensitive). If there is fluid on the joint or the blood tests are abnormal, a sample of joint fluid may be sent to the lab. This will be done in the office for knees, and by the radiologist with a flouoscope for hips. Occasionally bones scans may be ordered to look for loosening or failure of ingrowth but in a normal total joint the bone scan will often be abnormal for the first two years after surgery. An infection would require removal of the components followed by 6 weeks of intravenous antibiotics. Once the infection is resolved the total joint is then completed with permanent total joint components, unless other severe medical conditions prohibit it.

80% of patients with pain in their artificial joints do not have infections. Other causes tendonitis, inflammation of the lining of the knee, bleeding into the joint capsule, loosening of the implant, and instability of the joint. Most hip pain is actually from sciatica related to the back and in this instance further spine evaluation is recommended. Sciatica causes buttock pain. Hip pain is typically felt in the groin and front of the thigh. Revision surgery is usually only recommended for infection, loosening, fracture, component failure, severe or progressive osteolysis (dissolving bone around the implant), and significant instability (dislocations). Your surgeon should evaluate even a painless joint replacement one year after the surgery and then every 2 years to ensure that problems are caught early when surgical remedies are more successful.

**My therapist told me that my leg is longer. Do I need a lift in my shoe?**

While leg length may increase after surgery, the most common cause of leg length inequality, affecting 14% of patients undergoing hip replacement, is pelvic obliquity[84]. This can occur from contractures that form around the arthritic hip and from spine maladies, such as scoliosis. Symptoms usually abate after 6 months and are often treated with hula exercises and stretching to alleviate pelvic obliquity (tilted pelvis) and contractures. Using a heel lift for minor leg length discrepancies may slow the resolution of pelvic obliquity but still may be recommended to treat back pain in the setting of leg length inequality.

Major leg length discrepancies are more common in revision total joint procedures and in short women with short femoral necks and contractures about the hip. When total hip reconstruction results in over 2.7 cm of lengthening, injury to the nerves in the leg can occur and additional surgery may be recommended [115].

### When can I travel after joint replacement surgery?

Travel is reasonably safe after joint replacement while the blood is thinned with injectable blood thinners or warfarin. After discontinuing the use of blood thinners there is a period of time that you may be at risk for blood clots in the veins or clots that pass to the lungs (pulmonary emboli, which can be life-threatening). The risk can be decreased to some extent by taking breaks to stretch your legs. Especially common with air travel, blood clots can occur with any period of prolonged inactivity. Exercise and avoiding dehydration are recommended but the use of aspirin during travel is no longer recommended by the American College of Chest Surgeons to prevent clots in leg veins during travel. Should you need to travel soon after total joint replacement surgery, you should discuss the use of injectable blood thinners with your physician.

### What activities are not allowed after joint replacement?

Contact, court, and field sports have not been recommended, historically. Running is not recommended. Instead, you should consider substituting hiking, biking, cross country skiing, snowshoeing, moderate weight training and swimming. If you are an excellent skier, you may return to skiing conservatively. You must avoid falls to prevent fractures and dislocations around the total joint. The book is being rewritten on permissible sports for total joint patients. Tennis, for instance, looks like it may be safe [62].

Sexual relations are permissible as soon after surgery as participation is comfortable. The only precaution is against flexing the replaced hip into extreme positions, such as bringing the knees to the shoulders, as this may result in dislocation.

Kneeling may not ever become comfortable after knee replacement. About one-third of total knee patients cannot kneel comfortably and must use stools to garden or kneepads for activities that require kneeling. To regain the ability to kneel, try it first on your bed and then progress to the carpet, the grass, and finally on hard surfaces. If this is uncomfortable try using kneepads, such as those sold to gardeners.

Avoid extremes of position after hip replacement. Don't turn your knee inward to clip your toe nails or tie your shoes after hip replacement. Work between your legs when putting on shoes and socks or trimming your nails. Avoid all activities that flex the hip while rotating the knee inward. Don't bend over the side of your chair to pick up items that are on the floor. This is equivalent to rotating the hip inward because it rotates the pelvis outward over the fixed thigh. Usually you can flex your hip to 90 degrees for the first six weeks and work up to 120 degrees by 12 weeks but you should never force this motion. Sometimes motion is limited by one part of the artificial hip impinging on the other part and forcing more motion will lever the hip out of place. When lifting, bend at the knees and keep the operative side behind the nonoperative or less recently replaced side. Your physical or occupational therapist will be a good source of information about how to reengage in activities of daily living.

### When can I have the other knee/hip replaced?

Feel free to call your surgeon to schedule this whenever you feel physically and psychologically prepared. For most people this is about 6 weeks. If the first surgery resulted in a significant anemia, the second operation should be postponed until the blood is built back up with the use of iron and the passage of time. You should feel ready to proceed before you schedule the second operation. Many patients can avoid making a second insurance co-payment by timing their surgeries to occur in the same calendar year as they have already met their maximum co-pay for the year. Bear in mind, however, that this places a run on surgical services in the months of November and December. Usually, preoperative cardiac and pulmonary clearances are good for one year unless you have a change in your symptoms. A good guide that you may not need more detailed testing is the ability to do heavy housework and climb a flight of stairs without symptoms. This is just a guideline, however, and you will be rescheduled to see the appropriate specialist if there is any doubt.

# Chapter VIII. Revision Surgery

When a surgeon removes a total joint implant and replaces it with a new one, this is referred to as a "revision." The most common reasons for removal of an implant are loosening, infection, instability, severe wear, dissolving bone around the implant (osteolysis), breakage of the component, or a fracture of the bone around the implant. Occasionally, when the wear or instability of an artificial joint requires revision and the original implant is well fixed and of high quality only the bearing surface will need to be replaced. This is a more common situation in total hips that have been done in recent years as these modern implants were designed to allow this.

The revision implants almost always have longer stems to allow better bone fixation and support a more stable joint. They are much more expensive to the hospital than usual total joint implants. Revision hips are rarely cemented into place because short stemmed cemented revision hips have been shown to be less durable and the cement around long-stemmed cemented implants can be nearly impossible to remove.

Patients who have failed to follow up with their surgeons for many years after joint replacement may first be aware of a problem when there is already devastating loss of bone and this makes revision surgery much more complicated. Large segments of bone may have to be grafted into place or replaced with a mega prosthesis, which substitutes metal for structural bone. The results of revision are highly dependent on the skill of the surgeon, and are often worse than primary joint replacement. However, the best results may parallel the results of primary joint replacement and often there is less pain with revision surgery than with primary joint replacement.

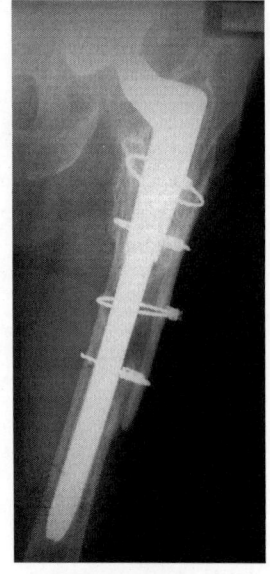

Revision surgery often requires removal and reattachment of parts of the bones, called "osteotomy." These procedures are actually done to avoid damage to the remaining bone and tendons during removal of the old total joint implants. They usually don't require any modification of therapy or loss of ability to bear weight on the new joint after the surgery. An osteotomy requires about 12 weeks of healing to unite with surrounding bone. Although uncommon, an osteotomy can fail to heal or fracture and lead to more surgery or more disability.

Bone from a bone bank can be grafted into areas of bone deficiency and this donor bone usually heals to your bone within 1-2 years. However, massive segments of bone are never fully incorporated. There is a small risk of bacterial or viral infection from this bone. Currently, the risk of AIDS from bone and tendon grafts approaches zero. The risk of bacterial infection is low and sporadic and should approach zero as tissue banks have implemented new standards for the harvesting and storage of donor tissues.

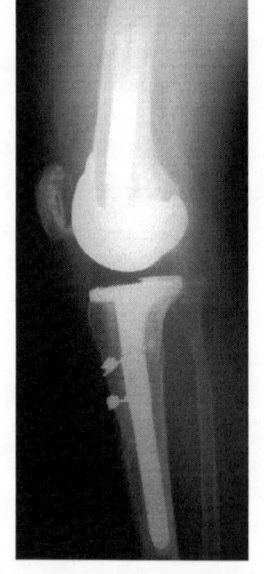

Revision surgery often requires much more creativity on the surgeon's part and it may be impossible for your surgeon to accurately predict the scope of the operation and the materials to be used before surgery. Similarly you won't know how much weight you can place on the operative leg, the possible need for bracing, or the precise nature of your rehabilitation until after the surgery. An ideal surgical result would not require bracing or excessive concern about placing weight on the leg.

Revision surgery is complicated economically. The hospitals performing this surgery don't come close to recouping their expenses from Medicare. The reimbursement for all of the hospital's expenses from Medicare usually doesn't even cover their cost for the total joint implants. Bone used for grafting is also very expensive. Our hospitals allow surgeons to do this surgery as a service to their patients. This is especially true in infection surgery, which often requires two operations and the prolonged use of expensive antibiotics.

Cup removed with worn out liner and metal debris in bone

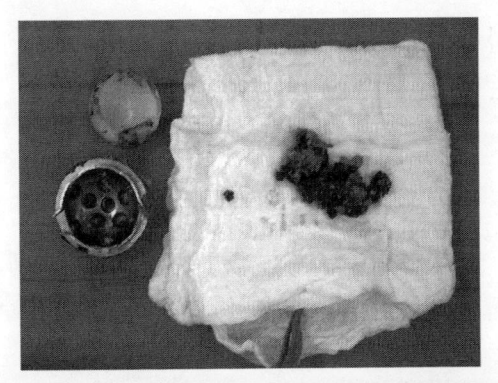

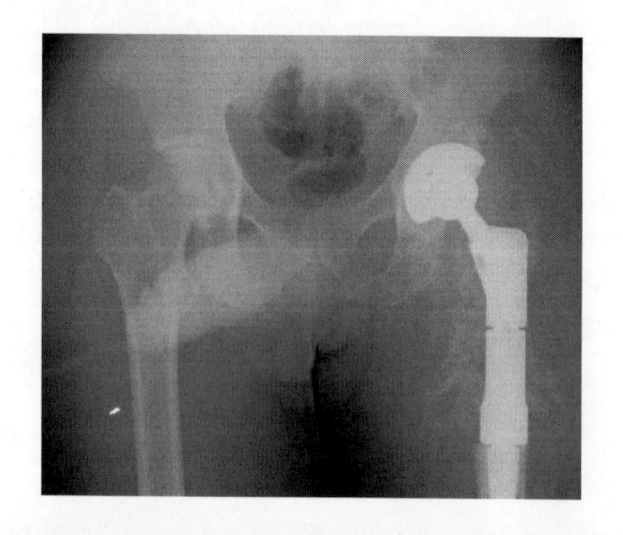

Megaprosthesis for deficient bone around hip (above).
Megaprosthesis for deficient bone around knee (below).

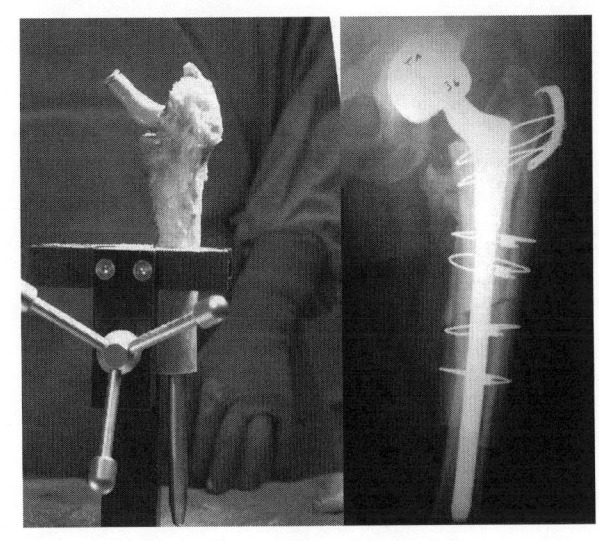

Massive loss of bone may require massive bone grafts (called "massive allografts")

Occasionally a severe fracture or tumor may require using revision components to replace the knee. In this instance, a hinged (RHK™, Zimmer) knee replacement was placed into a Trabecular Metal Cone™. The fracture had not healed for one year and the knee was completely unstable.

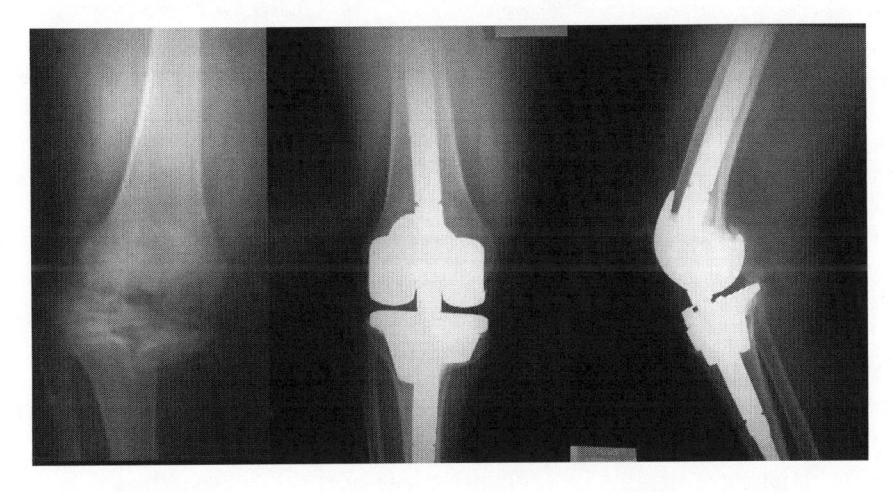

## Table 6. Risks of Revision Surgery

Data from Medicare have been collected for a one year periods of time and are included in the tables below. This includes complications which occurred in a 60 day period of time following surgery.

| Selected Major Risks of Revision Knee Replacement in Medicare Patients Who Underwent Surgery in the Year 2000 [1] | Revision in 11,726 Patients |
|---|---|
| Death | 1.1% |
| Readmission for additional knee surgery | 4.7% |
| Pulmonary Embolism | 0.5% |
| Wound infection | 1.8% |
| Heart Attack (MI) | 1.0% |
| Pneumonia requiring hospitalization | 1.4% |
| Knee manipulation with anesthesia | 1.3% |

| Selected Major Risks of Revision Hip Replacement in Medicare Patients Who Underwent Surgery between 7/1/95 and 6/30/96 [2] | Revision Surgery in 13,483 Patients |
|---|---|
| Death | 2.6% |
| Readmission to hospital | 10.0% |
| Pulmonary Embolism | 0.79% |
| Wound infection | 0.95% |
| Hip dislocation | 8.4% |

More complete follow-up is included in the Mayo Clinic's series. In this series, infection occurred in 3.2% of revision hip procedures and in 5.6% of revision knee procedures [54]. Risks are further enumerated in the sections on hip and knee replacement, infection surgery, and fracture surgery. Risks are higher in revision surgery than in first time procedures.

# Chapter IX. Fractures around Total Joint Implants

Fractures to the bones around a replaced joint, called "periprosthetic fractures," are difficult to treat and often result in the premature removal of the joint replacement implants. These fractures are more common in elderly patients with osteoporosis (brittle bone). They are also common in young, athletic patients who continue sports and recreational activities that place them at risk for high energy trauma.

This is an x-ray of the hip of a young woman from a ski area community who had a fracture from the insertion of the component at the time of surgery. This was successfully revised but soon after she fell on the ice and broke the bone beneath the stem of the femoral implant. This was treated with a plate and a combination of screws and cables (picture on left).

This repair failed and she was referred for yet another surgery. This surgery required removal of all of the components and surgical exposure of the entire thigh for placement of bone graft struts around the fracture site. A very long stem was used (right).

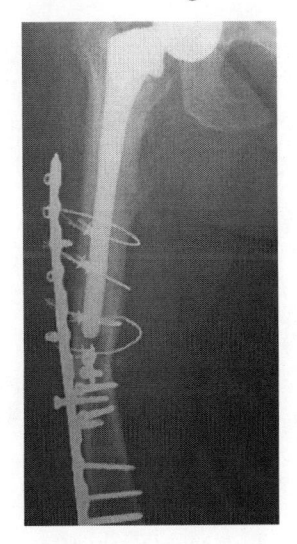

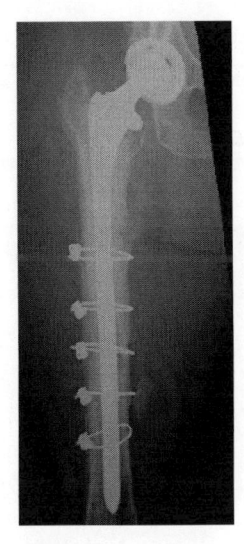

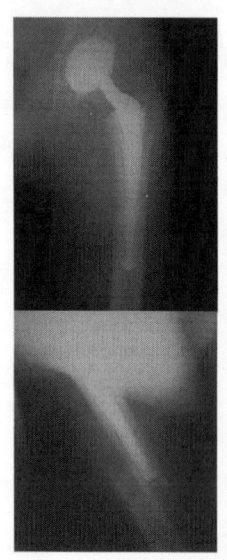

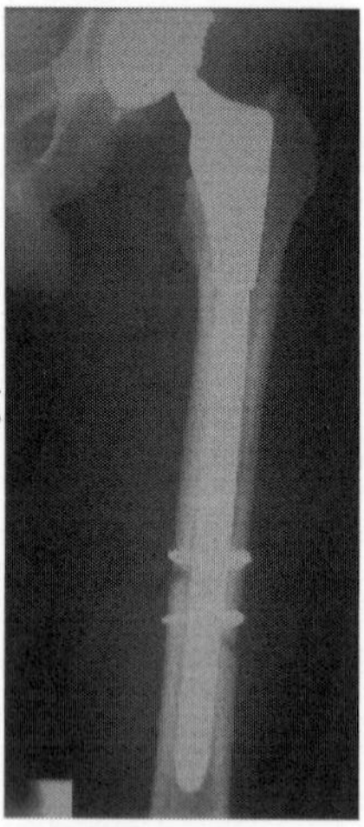

These x-rays are from a man in his late forties who was doing well after his hip was replaced (upper left). A month after his surgery, he had a serious mountain biking accident in which he broke the bone at the tip of the implant (lower left).

The stem of the hip replacement was then removed and replaced with a long stemmed implant. No bone grafting was required in this instance because his bone quality was high and the long stem provided strong fixation. The surgery was extensive, however. An incision through his thigh muscles was required to revise his hip and fix the fracture. The surgery resulted in leg length inequality.

This type of fracture can be common soon after hip replacement due to bone resorption from the surgery. His primary hip replacement was undone by this accident after only one month of service.

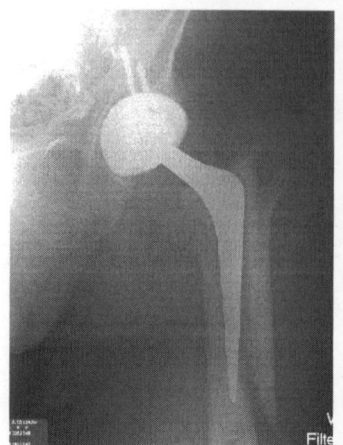

The films shown below are from an 87 year old man with kidney failure who fell off of a curb. The stem had been cemented into place, complicating this revision and fracture repair. This required the removal of all of the bone cement prior to placement of the long femoral stem. The broken bone was then fixed with strut grafts and cables as in the case above. This procedure is very extensive and the period of rehabilitation and healing is prolonged after such a surgery. Coexisting medical problems place this patient at increased risks for failure to heal the fracture (called "nonunion") and for infection.

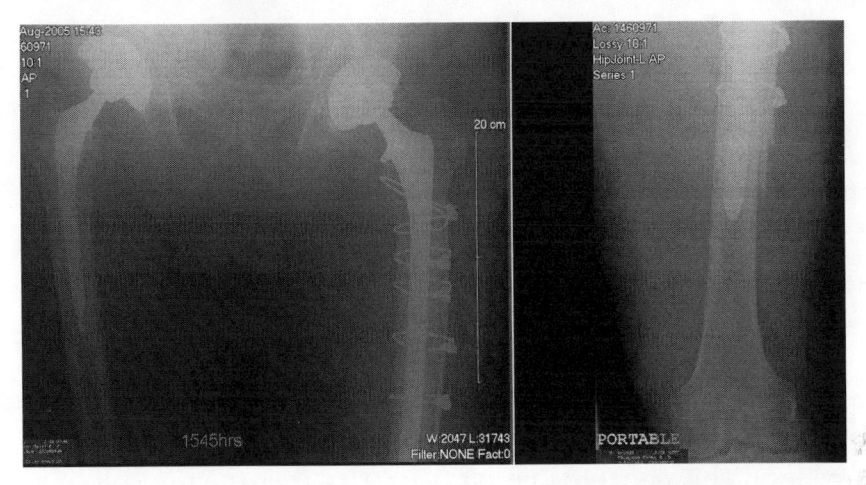

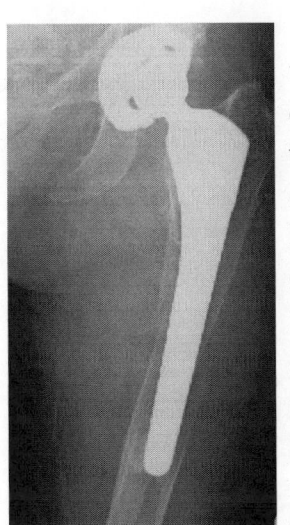

Sometimes the total hip components can be retained. This was the case for this elderly woman who had an ingrown femoral stem and a well-functioning hip when she fell. This fracture was fixed with a cable plate and bone grafting with strut grafts. This operation was still extensive and the fixation is less stable than the cases shown above. She was not allowed to bear weight on this leg until the fracture had healed (about 12 weeks).

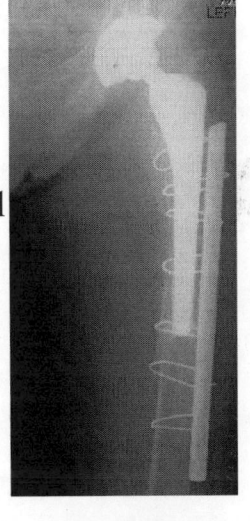

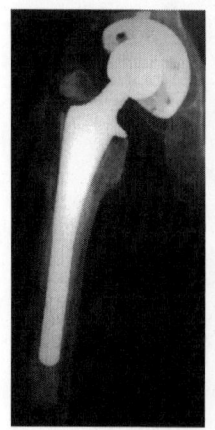

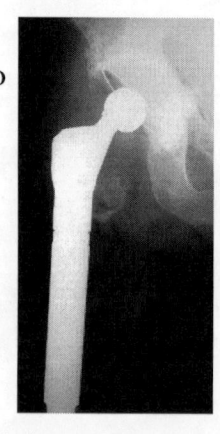

When the stem of the implant is loose and there is no bone left around the implant, the entire top of the bone may be replaced with an implant, called a "mega- prosthesis." This implant also had an infection which was treated with suppressive antibiotics before the fracture occurred.
This implant is more likely to dislocate than the usual type of hip replacement but may be the best option for very elderly patients who cannot keep weight off of the operated leg. This patient lived for two years after the top half of his femur was replaced with mega prosthesis. During that time he was able to walk comfortably and to live at home. His hip dislocated twice during that period.

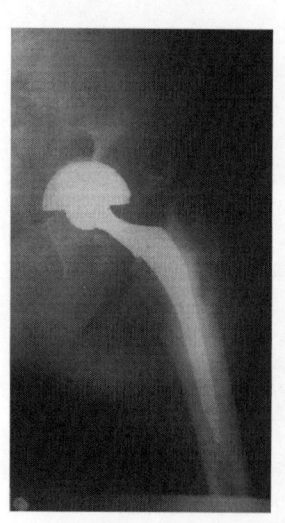

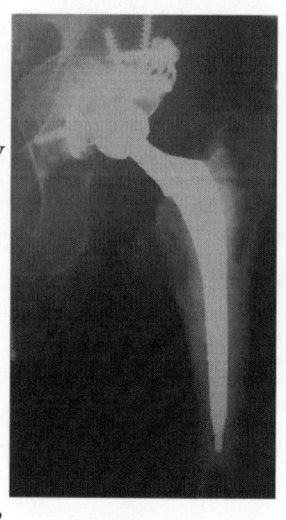

This severe pelvic fracture occurred soon after surgery and probably was the result of a fracture at the time of surgery that wasn't appreciated. Repair of this fracture required bone grafting and placement of a special apparatus called a "protrusio cage."
This late middle aged man continues to do well several years after repair of the fracture of the pelvis.

In the largest series of total hip implants with fractures at the tip of the stems, 90% of the surgeries lasted 5 years and just under 80% of the reconstructed hips lasted 10 years before they were lost to infection, got loose, or wore out [116]. These results are inferior to other joint results. We await a similar series for total knee periprosthetic fractures.

Fractures that occur after knee replacement are repaired without revision of the components in most instances. These x-rays are from a patient who was a middle aged rancher with osteoporosis. She had fractured her tibia before her knee was replaced. The long stem placed down this bone protected her from a re-fracture of her tibia, but the femur above the knee replacement was broken when she fell again. The bone was not deemed strong enough to hold an internal nail (rod) with locking screws.

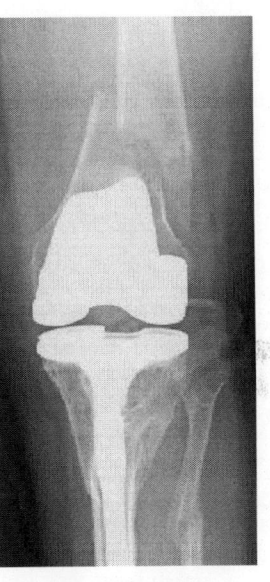

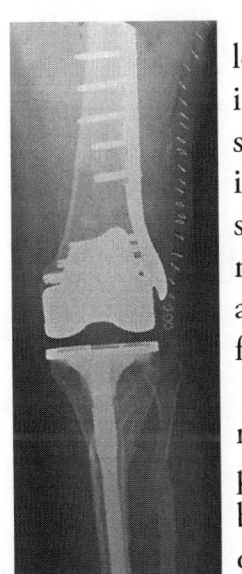

This fracture was treated with a long locking plate which also required an extensive incision through the muscles of the thigh, in spite of preoperative planning for minimally invasive surgery. The knee replacement was salvaged and did not require revision. The rehabilitation from this surgery, however, was at least as demanding as it would have been from revision surgery.

For fractures around a knee replacement, it requires less surgery to place a rod into the canal of the broken bone than it does to place a plate outside of the bone. There are three potential problems with fixation of fractures around knee replacements with rods, referred to as "intramedullary nailing." First, the knee replacement implants may block the point of entry for the rod (or intramedullary nail) to be placed.  Secondly, the bone may be broken or too weak to support the locking screws that attach the bone to the rod.  This causes the rod to loosen and protrude into the knee where it can damage the total knee implants.  Finally, the rod is not attached to the total knee implant and this can result in angulation of the fracture which can result in a bowlegged (varus) or knock-kneed (valgus) deformity of the knee, eventually causing the knee replacement to wear out prematurely.

Prevention of these fractures is the best treatment.  After total joint replacement surgery there is loss of bone around the implants.  Some of the loss occurs soon after the joint replacement surgery and the reasons for this are not certain [117].  That makes it important to avoid falls and high risk activities for several months after this type of surgery.  For middle aged and elderly patients with risk factors for brittle bones a scan for osteoporosis, such as a DEXA scan, should be performed every three years.  Bone loss discovered in screening tests should be treated aggressively with medications called "bisphosphonates", such as Fosemax™ and Actonel™, whenever possible.  These medications can actually add additional bone to the skeleton.  Calcium and vitamin D tend to just slow down loss of bone rather than adding new bone.

Falls in late middle aged patients should be prevented at all costs.  The use of bath mats, hand rails, and night lights as well as the removal of throw rugs from the home can help prevent such falls.  Older patients should avoid icy sidewalks and poorly lit or uneven terrain.

For younger patients, return to sports which result in falls should be postponed for at least three months after surgery. Activities that can result in collisions and falls from a height can result in fractures through healthy bone.  Total joint patients should respect the new threat that motorcycles, all-terrain vehicles, deer stands, farm animals, ladders, bicycles, and ski slopes pose to them.

# Chapter X.  Total Joint Infection Surgery

Infection rates vary from surgeon to surgeon and hospital to hospital. Reported total joint infection rates are variable but are often around 1%, and are more common after knee replacement than after hip replacement. Risks of infection are greatly increased if you had a previous infection in the replaced joint, when revision surgery is carried out, when surgery takes over four hours, and when your immune system is abnormal. Common causes of abnormal immune systems are rheumatoid arthritis, the use of immunosuppressive medications, kidney failure, liver failure, diabetes, malnutrition, AIDS, cancer, and obesity. Rheumatoid arthritis was associated with an infection rate 2.6 times greater than that seen in osteoarthritis in one study [120], about 4% [121]. Infection rates in diabetes have ranged from 0-7% in various studies [122, 123, 124]. Psoriasis was associated with increased infection rates in one study which found infections in 17% of patients with this skin disease [99]. Obesity is associated with poorly functioning white blood cells. Patients who have had organ transplants and must take immunosuppressive medications have an infection rate of 19%, and these infections are difficult to diagnose [33].

Previous septic arthritis (infection in the joint fluid) is associated with an infection rate of 4% [126]. Previous osteomyelitis (bone infection) is associated with an infection rate of 15% [126]. Prior knee surgery doubles the infection rate and previous hip surgery triples the infection rate [120]. An active infection elsewhere in the body is associated with higher infection rates to varying degrees [127].

Cirrhosis of the liver is associated with a 7% infection rate in knee replacement and joint replacement should be avoided in older patients with cirrhosis, when platelet counts are low, and when cirrhosis is associated with hepatitis B[29]. 13% of patients who were on dialysis for kidney failure developed infection in another study [30].

Obese patients had 6.7 times as many infections in total knee replacements and 4.2 times as many infections in total hip replacements in one study [128]. Patients with AIDS have a 14% rate of infection and patients who use intravenous drugs have a 25% infection rate [129]. Risk factors are summarized in the appendix.

There are bacteria in the air of the operating room at all times. Present in small enough numbers, the body's immune system takes care of them and infection does not occur. Infections from airborne bacteria in the operating room may originate from the staph bacteria that are harbored in the noses of the surgeon, operating team, or the patient themselves. While the use of ointments in the patient's nose can decrease the risk of transmitting these bacteria, we have seen the emergence of resistant bacteria in these circumstances and cannot recommend their use [130]. The length of your surgery may thus affect the chances of infection as may the operating room discipline, and the use of clean air procedures, such as space suits, laminar flow, and the maintenance of appropriate air flows and pressures in the operating room. Bacteria on the patient's skin (*staph. epidermidis*) is also an emerging source of resistant total joint infections but can be controlled to some extent with the use of special barrier drapes [131]. Shorter operating time is possibly the reason that surgeons and hospitals that do a lot of joint replacements have lower infection rates. Unless you have prohibitive allergies to multiple antibiotics, you should receive an antibiotic intravenously within one hour of your surgery. This will reduce your chance of infection by about 50%.

Rare causes of infection are contaminated implants, instruments, or bone cement. In January of 2003 one manufacturer recalled their bone cement due to concerns about the packaging of their product although the actual cement was sterile. In the 1980's a popular agent for skin preparation was recalled because the bacteria, pseudomonas, was growing in the solution used to prep the skin of the patients.  Even the antibotic most commonly used to prevent infections in total joint surgery, cefazolin, was recalled by Hanford Pharmaceuticals, Inc. due to bacterial contamination (*Bacillus pumilus, Staphylococcus hominis, Propionibacterium acnes, or Micrococcus luteus*) in 2006.

Late infections can occur from bacteria traveling through the bloodstream of patients.  For instance, the risk of a dental procedure causing a total joint infection is 1 per 10,000 from spread through the blood stream.  The vast majority of infections are not related to the incompetence of your surgeon, hospital, or equipment manufacturer.  Most infections are simply a matter of the small odds catching up with the occasional patient, in spite of very high quality surgical care.

Bacteria form biofilms, or slimes, around the implants that are infected.  Slimes are everywhere in nature.  They are the substances that make the inside of a hot tub or swimming pool slippery.  They cause dental plaque.  When algae form slimes it makes rocks in a stream slippery.  It is not possible at this time to remove these slimes from joint replacement implants in the body.  Bacteria cannot be eradicated from slimes with systemic or local antibiotics.  Work is underway to develop medications that dissolve biofilms and a chemical called chitosan which has promise to prevent biofilms from forming on implants in the first place[141].

The standard of care for infected total joints in the United States is to remove the artificial joint, possibly substituting a temporary one (PROSTELAC - prosthesis of antibiotic loaded acrylic cement) or a spacer, and to then replace durable components a few weeks later.  During these few weeks, you will be on intravenous antibiotics.  You may be asked to keep some or most of your weight off of the extremity to prevent fractures around a spacer.

For infections less than 3 days old and occasionally in other circumstances, a debridement, or a one stage reimplantation may be considered by your surgeon.  Stated simply, a debridement is a clean-out surgery. Although the success rates are not as high, it may be worth an attempt to eradicate the infection without multiple surgeries, spacers, or temporary implants. This decision is complex and must be made by your surgeon based on operative findings and general considerations about your overall health and well-being. Patients who have very recent onset of infection or have multiple other medical problems may be considered for a debridement with retention of the components that are solidly fixed. Infections less than three days old can occasionally be eradicated without removing the implants. On occasion, there may be uncertainty about the presence of deep infection and, in this instance, the components may be preserved.

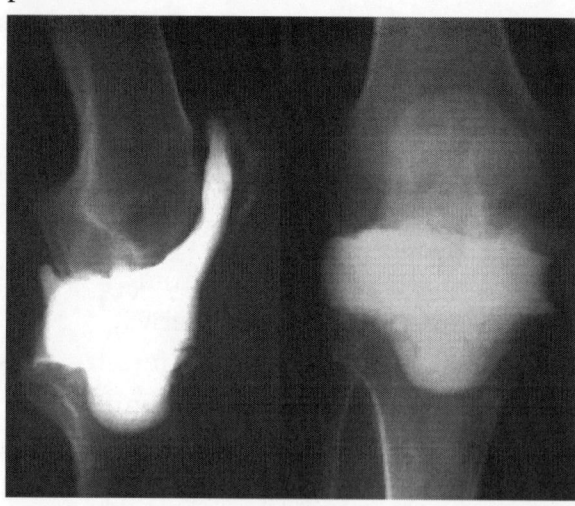

This is an example of an infected total knee after the components have been removed and an antibiotic spacer has been placed. It is not possible to move this knee until it has been converted back to a total knee replacement with removal of the spacer and revision surgery. If this knee is flexed, bone damage may occur. While this knee is at this intermediate stage, a brace should be worn and only touch toe weight bearing should be allowed to prevent fractures and additional bone loss.

The bone cement antibiotic spacers can also develop slimes around them. One study showed bacteria growing on 90% of the antibiotic beads at the time of removal [50]. This is especially true if too low of a dose of antibiotics is used in the cement. In this case, the emergence of bacterial resistance is common. Bone cement is usually hand mixed with 3 grams of antibiotics to ensure that enough antibiotic elutes from the spacer or beads to kill the surrounding bacteria, not just to let them develop resistance as seen in lower dosage preparations. The spacers are not necessarily sterile when they are removed.

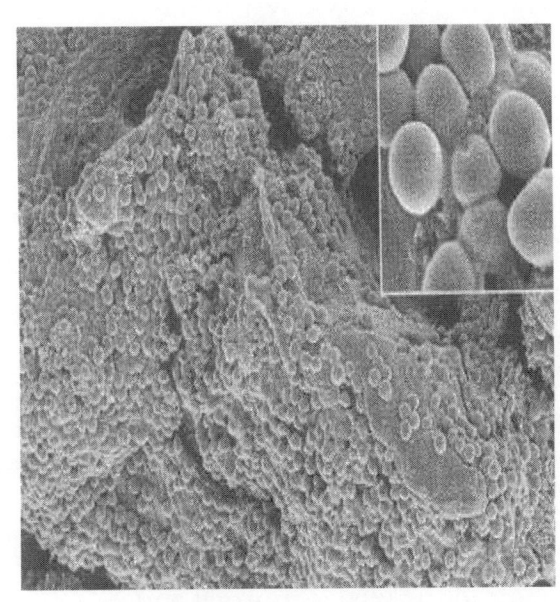

Scanning electron mircrography (a powerful microscope photo) of bacteria on PMMA beads used to treat an infection.
Courtesy of
Danielle Neut
University of Groningen
Groningen, Netherlands

There are two blood tests that are beneficial to your surgeon and infectious disease specialist as they judge the activity of your infection while you are under treatment for a total joint infection. Following successful infection surgery, the C-reactive protein (CRP) is the first lab value to return to normal, followed by the erythrocyte (red blood cell) sedimentation rate (ESR). Either of these tests may be abnormal for other reasons, so it is important to inform your treating physician if you have other symptoms (sinus infection, bladder infection, diarrhea, etc) while you are receiving antibiotics. Occasionally, super-infections emerge. They can be serious and require additional care. One type of diarrhea, from a type of bacteria named *clostridium difficile*, can be dangerous and hard to eradicate. You should notify your physicians if you develop diarrhea while on antibiotics.

Laboratory monitoring of some antibiotic levels is important because these antibiotics can damage your kidneys or your hearing. If, by 6 weeks, the CRP is not substantially improved, another debridement may be recommended prior to the definitive surgery to revise your total joint.

Another useful marker of infection is IL-6 (interleukin 6), which is expected to be available as a blood test in the future. Because blood levels of IL-6 return to normal within three days of surgery, this test may be of more use to surgeons than the C-reactive protein (CRP) and the erythrocyte sedimentation rate (ESR). Below is a table adapted from Di Cesare, et al. which reports the accuracy of various blood tests in discovering infections around total joint implants [132].

| Test for Infection | Sensitivity | Specificity | Accuracy |
|---|---|---|---|
| ESR | 1.00 | 0.56 | 69% |
| CRP | 0.94 | 0.78 | 83% |
| IL-6 | 1.00 | 0.95 | 97% |

When the new joint is placed, you may be asked to remain on an IV antibiotic until the CRP is normal and then an oral antibiotic until the ESR becomes normal, but many variations in antibiotic use after surgery are favored by different surgeons. The combination of surgery and antibiotic therapy will eradicate the infection about 80% of the time. Life long antibiotics are not recommended except when a patient is too fragile to survive surgery, when a patient has an impaired immune system, or when multiple attempts to cure an infection fail and for patients who will not accept a Girdlestone procedure (permanent removal of a hip joint), or an amputation or a fusion (in the case of the knee joint). A successful surgical treatment of infection, by definition, does not result in life-long antibiotic therapy. The antibiotics can be discontinued when the markers of infection in the blood serum (sedimentation rate and C-reactive protein) have normalized, unless extreme problems with the immune system predict failure of the usual care. At the point that antibiotics are stopped altogether, the labs are followed periodically. If these lab results again become worrisome, a decision about more surgery, or long-term suppressive antibiotics should be made between the surgeon, the infectious disease specialist, and the patient. Long-term use of antibiotics is believed to lead to the development of antibiotic resistance in the bacteria that are causing infections. Even our most bullet-proof antibiotic, vancomycin, can't eradicate some of these bacterial strains. Some of the drugs, used to treat these resistant bacterial infections, such as Zyvox™ and vancomycin, can cause immunosuppression from their toxic effects on the bone marrow.

Recurrent infections are too common, especially for patients who have so-called "host factors," which include: cancer, immunosuppression from diseases or immunosuppression from medications used to treat their disease, diabetes, organ failure, malnutrition, vascular disease, sickle cell disease, and severe obesity. Smoking is also a host factor which predicts a poor outcome. It may be the only host factor that a patient can control. We can salvage most of the infected joints with these procedures. The options left if infection recurs include the following:

1. Attempt another staged revision. This is preferable except in patients who have host factors that are too forbidding.
2. In the case of the hip, your surgeon can remove the joint. This is called a Girdlestone procedure and may result in an acceptable level of pain, and usually requires a built up shoe and results in a limp and the need for a walker or crutches.
3. In the case of the knee, fusion or an amputation above the knee may be considered. Fusions are not well accepted due to limp and problems sitting with a knee that doesn't bend. Patients with knee fusions

require special sitting arrangements on airlines and, because they can't bend their knee, often trip other patrons in movie theaters, but a fused knee is often painless. Attempts to fuse the knee are not always successful and may require more surgery and prolonged bracing. Function with a prosthetic leg after amputation may be preferable to the function with a fused knee for some patients. For this reason, amputation may be a better choice. It is hard for an older patient to learn how to walk with an artificial leg and walking requires more energy for all patients. Some older patients will wind up in a wheelchair after amputation surgery.

4. Suppressive antibiotics can be used for patients who are too debilitated even for an amputation or Girdlestone hip arthroplasty, or for patients who refuse further surgery. However, antibiotic resistance may emerge over time and result in failure to control the infection with antibiotics. Permanent use of antibiotics is the least preferable option for the treatment of a deep infection around a total joint implant.

# Chapter XI. A Primer on the Materials Used in Joint Replacement

## In Short:

Every so often, a patient will make an appointment with their orthopedic surgeon for a checkup on a hip or knee replacement done over 30 years previously. What made these joints so durable in these people wasn't that the technology was so good in the 1970's, but that the surgeon performed the surgery perfectly. A well done surgery can do more to extend the lifetime of your total joint implants than using the best technology available. What you need to know about the materials used for your surgery is whether the implants could last as long as you will live, or if they are likely to wear out. Your surgeon, of course, cannot make any guarantees, but you should ask him or her if there are ways to make the replacement more durable and if the hospital restricts the use of new technology based on the type of insurance that you have. You should know that the average 65 year-old women will live about 20 more years and that the average 65 year-old man will live about 15 more years and that these numbers are slowly increasing for patients who aren't overweight. A careful surgeon and a hospital without restrictions on the expense of the implants are a good, but not a foolproof combination. It is prudent, if you are elderly, not to take chances on new techniques and new technologies if you wouldn't expect to outlive the time-tested methods that have made joint replacement so successful in the past. New techniques and technologies sometimes fail. You don't want to wind up needing another surgery because you demanded a new type of total joint that didn't stand the test of time.

If you don't trust your surgeon to make the correct decision about the surgical technique and type of implant to be used, find another surgeon. If you don't trust your hospital to make decisions about your well-being that are not purely financial decisions, find another hospital.

The people who need to think about taking the small risks associated with very new technologies are those who expect to live for more than twenty years based on their age, their health, and their family history. This chapter is a guide for these patients.

## The Details:

The first problems that we faced with the advent of total joint replacement surgery were implant loosening and breakage of the implants. The original materials used in total joint replacement were stainless steel and polyethylene, cemented into place. Breakage was avoided by changes in metallurgy, design changes, and changes in surgical technique. While early designs were often durable, the implants would eventually become loose and have to be revised. This was especially true for knee replacements and the cup in hip replacements. Materials for hip replacements are now often designed with the intent of obtaining direct bone ingrowth, particularly between the cup and the pelvis. However, cemented stems are still relatively durable, especially polished stems. The use of bone ingrowth in knee replacements has not always made the knee more durable, perhaps less so [63]. Efforts to make knee replacements more durable have focused on knee geometry and implant bearing surfaces.

Wear particles have been the focus of the most recent innovations in total joint materials. In the hip, the types of wear that predominate are called abrasive wear and adhesive wear which produce small particles that can incite bone resorption, called osteolysis which literally means "dissolving bone." In the knee, shearing forces predominate and the predominant types of wear are pitting and delamination.

In the hip, with metal on polyethylene designs, it was clear that patients with less than 0.2mm of wear per year were much less affected by loosening and bone loss. Many of the young, active patients had more wear and their hip replacemnts were not as durable due to wear and osteolysis (dissolving bone). Osteolysis was found to be caused by small wear particles absorbed by scavenger cells, called macrophages. These cells secrete hormones that stimulate bone resorption by cells called osteoclasts, or bone destroyers. While work is underway to test drugs that slow down this bone resorption, implant manufacturing has necessarily focused on decreasing the wear in bearing surfaces. Large amounts of wear particles that can be absorbed by scavenger cells are simply unacceptable.

In the knee, work has be focused on decreasing contact stresses and shearing forces with the use of rotating platform knees, changes in the congruency of the implants, the use of ceramics and improvements in polyethylene. The largest improvements were made by changing the sterilization process so that the plastic was no longer degraded from oxidation which occurred from sterilization with irradiation in air.

Any material, such as bone, cement, metal, ceramic, or polyethylene can form a wear particle. Present in large enough quantities, and in a particular size range, the wear particles will cause osteolysis. We have successfully decreased the amount of wear particles with the use of highly cross-linked polyethylene (plastic), ceramics, and the use of all-metal articulations. Some of these articulations produce much smaller particles than we have previously studied in osteolysis research. The very small particles are under study in animals at the time of this writing.

We have developed implants that show a marked reduction in the amount of osteolysis in hip replacements with the use of these new bearing surfaces, called alternative bearing surfaces. With both the reduction of loosening seen in the old cemented cups and the reduction of osteolysis with new alternative bearing surfaces, we are now sufficiently confident that a hip replacement will be durable enough to recommend this surgery to younger, more active patients.

Because most of the particles from the bearing surfaces of knee implants are too large to be engulfed by scavenger cells, the knee is less affected by osteolysis than the hip. In the knee, wear particles are more likely to cause swelling and or stretching of the ligaments over time from inflammation of the cells that line the knee. Historically, the small wear particles in the hip caused osteolysis, and the larger wear particles in the knee caused synovitis, inflammation of the soft tissues around the joint. The solution to these two very different problems has required different sets of technologies.

Ceramics in orthopedic use are compounds of metals and oxygen combined with trace amounts of other nonmetallic substances to improve their mechanical characteristics. Alumina is produced from aluminum, and zirconia is produced from another metal, zirconium. Ceramic compounds have more ions (electrically charged particles) on their surfaces that help to introduce a layer of joint fluid between the components and decrease friction in an artificial joint. They are harder than their associated metals and produce less wear on the surfaces adjacent to them because they are more resistant to the development of surface scratches. Ceramics were first introduced to hip replacement surgery to prevent osteolysis. All-metal hips, ceramics, and highly cross-linked polyethylene have found their way into our hip implants to prevent dissolving bone. Each of these technologies seems quite promising. A durable ceramic implant for the knee, oxidized zirconium (Oxinium™), was only recently developed. The decision to use each of these newer technologies has to be individualized until we have long-term experience with them.

One type of ceramic used in hips, zirconia, was recalled in 2001 by multiple manufacturers due to the risk of fracture from lots that were not manufactured to the usual specifications. Two batches of ceramic heads manufactured in France broke with frequencies of 15% and 3.4% respectively, and accounted for a total of 1200 prosthetic heads recalled. The risk of breakage from properly manufactured ceramic components is 0.04% and should not deter us from using ceramics for younger, more active patients now that we are again confident in the manufacturing process. The other type of ceramic in wide use, alumina, was originally thought by most to be inferior to zirconia, but has not had any recalls for defects in manufacturing. Refinements in alumina have now led to equivalent properties in zirconia, and both are now widely used. Ceramic-on-ceramic hips are made of alumina. The newest ceramic actually combines small amounts of zirconia and yttrium with alumina to produce a stronger matrix. This ceramic, marketed as Biolox-delta™ has been manufactured for all ceramic hip replacement and solid ceramic knee replacements.

The implantation of the all-ceramic hip requires exacting surgical technique so that they will not chip, dislocate, or fracture. With perfect surgical technique engineers feel that this bearing surface could last the lifetime of our youngest patients, but this claim cannot be validated without long-term follow up from large numbers of patients and surgeons. Results from solid ceramic knee replacements are not yet known.

Surface ceramic knees and hip heads are made of zirconium with the surface converted to zirconia, its ceramic derivative. Surface ceramic knees and hips are as resistant to fractures as metals and are very resistant to small scratches like those that have been found in older metal joint implants after years of use. Even when used with highly cross-linked polyethylene, scratches may greatly reduce the longevity of the implant. The metal underneath the surface ceramic material is not an ideal surface for bone ingrowth, and this lead to a voluntary recall on the surface ceramic ingrowth knee prosthesis when the manufacturer, Smith and Nephew, discovered ingrowth failure in a number of the devices implanted without cement. Failure of bone ingrowth is not rare in traditional metal implants which are still on the market. Many surgeons feel that the results of all knee replacements are better if all components are cemented [63], so this recall did not affect the patients of these surgeons.

Refinements have been made to the plastic substance used in knee and hip replacements, including improvements in their chemical composition, improvements in the locking mechanisms that affix this plastic to its titanium backing, polishing the components adjacent to the plastic to reduce wear where it is not supposed to occur, and improvements in the sterilization process to eliminate oxidation of this polyethylene plastic. Heat pressed components are believed to be inferior to compression molded components. While there is enthusiasm to import the cross-linking technology used in hips to knee implants, the knee has different wear patterns and strength requirements. The use of highly cross-linked polyethylene as a bearing surface appears safe and promising to increase the durability of hip replacements. Cross-linking was originally felt to be unsafe for knee replacements because of risks of breakage. Early cross linking and melting technology resulted in weaker plastic. Newer cross-linking technologies, such as the x3 highly cross-linked polyethylene available from Stryker™, show the same stregth characteristics as non cross-linked polyethylene and are now available for use in knee replacements. Long-term follow-up will be required to determine whether highly crosslinked polyethylene will make knee replacements more durable.

The metals used in joint replacement implants, and the use of advanced metallurgical techniques to strengthen them, have nearly eliminated their failure. The parts of joints exposed to motion are rarely made of stainless steel or titanium anymore. These metals may not be as good at joint surfaces due to small surface imperfections (pitting in stainless steel) and titanium's tendency to burnish with wear. The all-metal hips, metal hip heads, and metal knee motion surfaces are now usually made of a cobalt-chrome-molybdenum alloy.

The parts where motion is not supposed to occur and the parts in which bone ingrowth occurs are made of titanium in most instances. Tantalum, another elemental metal, has been used for bone and soft tissue ingrowth with promising results, in a material marketed by Zimmer as Trabecular Metal™.

All-metal hip replacements have been associated with measurable metal levels in the blood and urine of patients after surgery. Researchers have not discovered any diseases associated with this unusual metal exposure, but surgeons are concerned about this finding, especially for our patients with metal allergies. All metal hips have been associated with abnormal immune cell proliferation around blood vessels in the tissues around the hip that are not seen in metal on polyethylene hips [93]. All metal hips have been associated with the possible development of metal allergies[97]. Also, gene mutations have been seen more frequently in the blood (lymphocytes) after all-metal hip replacement [98]. This finding had previously only been reported in failed metal on polyethylene hips. All-metal hips have been used for many years in small numbers with much less tendency to cause dissolving bone (osteolysis) than the early polyethylene components that they competed against.

| | | | |
|---|---|---|---|
| **Ceramic on Ceramic (this technology is available for hip replacement, but not knee replacement)**  Wear rate 0-2.5µ per component per year. | 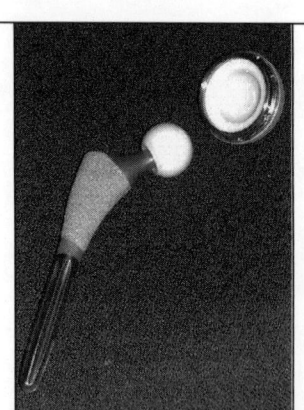 | There are very small particles (nanometer) in this material, and very little wear compared to conventional materials. Dissolving bone (osteolysis) is rare unless one of the components becomes loose. Many believe that this hip may last a lifetime. | Ceramics can chip or break (4 of every 10,000). Debris can't be fully removed after breakage, compromising the results of revision (redo) surgery. Ceramic cups must be perfectly placed to prevent chipping, breakage or dislocation. Elevated liners (which reduce dislocations) are not available. |
| **Surface Ceramic (Oxinium™) on Polyethylene**  Wear rate inferred from hip data 0-150µ per year; surface ceramics in knees are under study. | 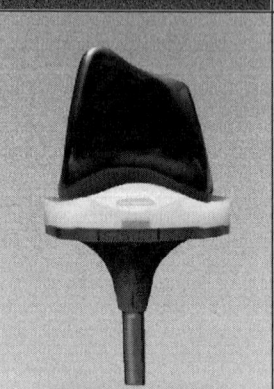 | Fracture-proof ceramic (Oxinium™) will not break and is available for both hip and knee replacements. Surface ceramic is much more resistant to scratches that occur from normal wear in the body than are metals. Fewer scratches results in less polyethylene wear. | Very low wear but small particles (0.2 micron). Thin layer of ceramic is actually thicker than the polished layer of metal implants (5 microns versus 2 microns) but doesn't self polish like metals. Surface ceramics do not allow cementless knee replacement or all-ceramic hip replacement. |
| **Surface Ceramic (Oxinium™) on Highly Cross-Linked Polyethylene**  Wear rates inferred from solid ceramic on poly hip data 0-150µ per year. | 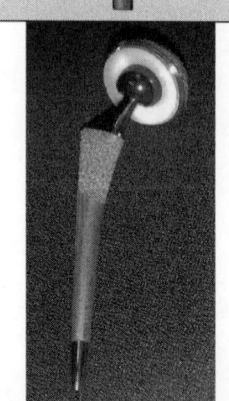 | Highly cross lined poly reduces wear especially if protected from surface scratches with ceramic heads. Larger heads, elevated liners, oblique liners, and constrained liners can be used to reduce impingement and dislocations without excessive wear. | More cross-linking means less wear but more tendencies toward breakage. This is well tolerated in hips with metal backed cups but cross-linking has to be reduced in knee replacements. |
| **Metal-on-metal**  Wear rate 27.8 µ in the first year, 6.2 µ subsequent years [133]. | 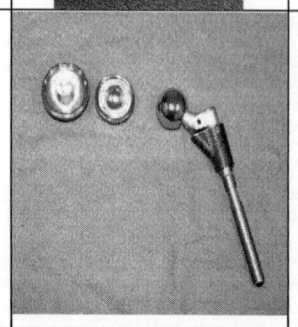 | Very little wear and very low rates of osteolysis. These implants have been used in small numbers since the 1930's although designs for these implants have been greatly improved over recent years. These implants should be the most resistant to breakage. | Metal levels can be detected in the blood and urine after an all-metal hip replacement, which is worrisome but has not been associated with any systemic disease. Metal in the blood stream and lymphatic system is particularly worrisome for patients with metal allergies. This implant system is not yet available with elevated liners to reduce dislocations. |

# Appendix

## Table 7. Anti-inflammatory medications used for the treatment of osteoarthritis (OA)

| Chemical Name | Brand Name | Dose for OA in young adults, normal size & renal function |
|---|---|---|
| Celecoxib | Celebrex | 200mg once daily |
| Diclofenac | Cataflam, Voltaren, Arthrotec | 50-75mg twice daily (200mg misoporstol in Arthrotec) |
| Diflunisal | Dolobid | 250-500mg twice daily |
| Etodolac | Lodine, Lodine XL | 200-400 mg three times daily |
| Fenoprofen | Nalfon, Nalfon 200 | 300-600mg three times daily |
| Flurbiprofen | Ansaid | 100mg three times daily |
| Ibuprofen | Motrin and many others. | 200-800mg three times daily |
| Indomethacin | Indocin, Indocin SR, Indo-Lemmon, Indomethegan | 25-50 mg 2-4 times daily / 50-75mg twice daily for extended release form |
| Ketoprofen | Oruvail, Orudis, Actron | 50mg three times daily / 150mg daily in extended release |
| Meloxicam | Mobic | 7.5-15mg one daily |
| Nabumetone | Relafen | 500mg twice daily |
| Naproxen** | Aleve, Naprosyn, Anaprox, and others. | 220-500mg twice daily |
| Oxaprozin | Daypro | 600-1200mg once daily |
| Piroxicam | Feldene | 10-20mg once daily |
| Salsalate | Disalcid | 1000mg three times daily |
| Sulindac | Clinoril | 150-200mg twice daily |
| Tolmetin | Tolectin, Tolectin DS, Tolectin 600 | 400mg three times daily |

## Table 8. When to Stop Arthritis Medications and Immunosuppressive Medications before Surgery (adapted from Rosandich PA, Kelley JT, Conn DL) [35]

| Drug | Brand Name | Class of Medication | Recommendations |
|---|---|---|---|
| Aspirin | | Nonselective NSAID | Stop 7-10 days before surgery and resume when warfarin (Coumadin™) or LMW heparin (Lovenox™) is stopped. |
| Salicylates | Disalcid™, multiple others | Non-acetylated salicylates | No need to stop before surgery. Blood platelets work normally. |
| Traditional NSAIDs | See table 2 | Nonselective NSAID | Stop 7 days before surgery (held at least 5 times longer than the half life of the medicine). May block bone ingrowth after cementless joint replacement. |
| COX-2 NSAIDs | Celebrex™ | COX-2 selective NSAID | Does not have to be stopped prior to surgery. |
| Glucocorticoids | Prednisone | Steroid Anti-inflammatory | Should not be stopped around the time of surgery and is given IV before surgery to prevent circulatory problems from stress (hydrocortisone 100mg). |
| Methotrexate | Amethopterin™ Folex™ Folex PFS™ Rheumatrex™ Trexall™ | Folic acid antimetabolite used in chemotherapy | Stop one week before surgery and restart one week after. Continuation of this medication around the time of surgery is more dangerous for elderly patients and those with kidney or liver disease. |
| Leflunomide | Arava™ | Pyrimidine antimetabolite | Stop 3 weeks prior to surgery and restart 3 days after surgery. |
| Sulfasalazine | Azulfidine™ | Local anti-inflammatory in the colon | Stop one day prior to surgery and restart 3 days after surgery. |
| Azathioprine | Imuran™ Azasan™ | Purine antimetabolite inhibits DNA and RNA synthesis | Stop one day prior to surgery and restart 3 days after surgery, or as instructed by your transplant specialist. |
| Hydroxychloroquine | Plaquenil™ | Antimalarial and anti-inflammatory | No need to stop. |
| Infliximab | Remicade™ | TNF-inhibitor $t_{½}$ = 8-9½ days | Stop one week before surgery and restart 1-2 weeks later. |
| Etanercept | Enbrel™ | TNF-inhibitor $t_{½}$ = 4½ days | Stop one week before surgery and restart 1-2 weeks later. |
| Adalimumab | Humira™ | TNF-inhibitor | Stop one week before surgery and restart 1-2 weeks later. |
| Anakinra | Kineret™ | IL-1 antagonist | Stop one week before surgery and restart 1-2 weeks later. |
| Rituximab | Rituxan™ | Anti-CD20 monoclonal antibody | Stop at least one week before surgery and restart 1-2 weeks later. |

## Table 9. Comparative Risks of Blood Transfusion

**Blood Risks per Pint**

| Complication | Risk per Unit (pint) |
|---|---|
| Minor Allergic Reaction | |
| Bacterial Infection | 1:2500 |
| Viral Hepatitis | 1:5000 |
| Lung Injury | 1:5000 |
| Hemolytic Reaction (Red cells break up) | 1:6000 |
| Hepatitis B | 1:205,000 |
| Hepatitis C | 1:935,000 |
| HIV/AIDS | 1:2,135,000 |
| Anaphylaxis (Major Allergic Reaction) | 1:500,000 |
| Fatal hemolytic Reaction | 1:600,000 |
| HTLV I/II infection (Risk for bone marrow cancers) | 1:641,000 |
| CVHD | Rare |
| Immunomodulation (lowering resistance to infection) | Unknown |

**Lifetime Risks of Death Faced Everyday**

| Possible Causes of Death | Lifetime Risk |
|---|---|
| Death from smoking one pack per day | 1:200 |
| Death from influenza | 1:5000 |
| Death from an auto accident | 1:6000 |
| Death from a plane accident (frequent flyer) | 1:20,000 |
| Death from leukemia | 1:50,000 |
| Death from the use of birth control pills | 1:50,000 |
| Death from tornadoes if you live in the Midwest | 1:445,000 |
| Death from a flood | 1:445,000 |
| Death from an earthquake if you live in California | 1:558,000 |

Adapted from John J. Callaghan, MD, "Blood Management in THR," Current Concepts in Joint Replacement Spring, 2000, Las Vegas, NV

## Table 10. Health Information
**CONFIDENTIAL**  (Fill out before doctors appointment)
Date of Visit_____

NAME_____ AGE_____ DOB_____
PHONE NUMBERS_____ SSN_____
INSURER_____ INS/GROUP NUMBERS_____
PRIMARY PHYSICIAN_____ REFERRED BY_____
REASON FOR EVALUATION_____

Past Medical History - Review of Systems (circle or underline all that apply)

**CANCER** – TYPE_____ PREVIOUS CHEMOTHERAPY, PREVIOUS CANCER SURGERY, CURRENTLY FREE OF DISEASE (Y/N), FOR HOW MANY YEARS? _____, PREVIOUS RADIATION THERAPY (BODY REGIONS_____, WHEN?_____), SKIN CANCERS: (MELANOMA Y/N)

**MOUTH/THROAT/AIRWAY** – INFECTIONS, LOOSE TEETH, DENTURES, GUM DISEASE, HAVE ANESTHESIA PROVIDERS EVER HAD PROBLEMS WITH INTUBATION? (Y/N)_____

**HEART/VASCULAR** – PREVIOUS HEART ATTACK (AT WHAT AGE?_____), HEART RHYTHM PROBLEMS, PACEMAKER, FAINTING, CHEST DISCOMFORT ON EXERTION, PREVIOUS HEART SURGERY, ANGIOPLASTY OR STENTS (WHAT & WHEN?_____), MURMUR, HEART FAILURE, SIT UP AT NIGHT TO CATCH YOUR BREATH, SLEEP ON MULTIPLE PILLOWS, CALF PAIN WITH WALKING, POOR BLOOD SUPPLY TO HANDS OR FEET, ARE YOU ABLE TO WALK UP A FLIGHT OF STAIRS? Y/N   CAN YOU DO HEAVY HOUSEWORK? Y/N

**ENDOCRINE** – OSTEOPEROSIS, STEROIDS (PREDNISONE), THYROID PROBLEMS, DIABETES, USE INSULIN? Y/N

**LUNGS** - HISTORY OF PULMONARY EMBOLUS (WHEN?_____), TB, ASTHMA, EMPHYSEMA, DO YOU USE OXYGEN AT HOME? (Y/N), HOW MUCH? _____, USE INHALERS? (Y/N), DO YOU SNORE LOUDLY, STOP BREATHING, HAVE SLEEP APNEA, STILL FEEL TIRED AFTER SLEEP?

**BLOOD** - BLEEDING TENDENCIES/BLOOD THINNERS, BLOOD CLOTS/PHLEBITIS, JEHOVAH'S WITNESS

**GI** - PEPTIC ULCER DISEASE/ULCERS, BLEEDING, REFLUX/HIATAL HERNIA, HEPATITIS (A/B/C), LIVER DISEASE, PANCREATITIS, PROBLEMS SWALLOWING, SPECIAL DIET_____

**KIDNEYS** - HYPERTENSION, KIDNEY STONES, KIDNEY DISEASE, DIALYSIS (DAYS OF DIALYSIS_____), FREQUENT INFECTIONS, PROSTATE PROBLEMS, PREVIOUS SURGERY_____, CATHETER USE, FREQUENCY OF URINATION, URGENCY, HOW MANY TIMES NIGHTLY DO YOU URINATE?_____

**NEUROLOGICAL** – STROKE, SEIZURES, PRIOR BRAIN OR SPINE SURGERY (WHAT/WHEN?_____), MUSCLE DISEASE, SCIATICA, HEADACHES, FORGETFULNESS, ALTZHEIMER'S DISEASE, TREMOR, PARKINSON'S DISEASE, DEPRESSION, BIPOLAR DISORDER, SCHIZOPHRENIA, ATTENTION DEFICIT DISORDER, OBSESSIVE-COMPULSIVE DISORDER, ANXIETY DISORDER/PANIC ATTACKS, CLAUSTROPHOBIA

**INFECTIOUS DISEASE** – FREQUENT INFECTIONS, IMMUNE SYSTEM DISORDERS (HIV, AIDS), POSITIVE SKIN TEST FOR TB (Y/N), PREVIOUS BONE OR JOINT INFECTIONS, ACTIVE INFECTIONS (LIST_____), DO YOU TAKE ENBREL, REMICADE, IMMURAN, METHOTREXATE, OR PREDNISONE?

**RHEUMATOLOGY** – RHEUMATOID ARTHRITIS (HAVE YOU HAD TENDON RUPTURE, NECK PAIN, OR GET ELECTRIC SENSATIONS IN ARMS OR LEGS WITH NECK MOTION?), LUPUS, GOUT, OTHER_____

**WOMEN:** MENOPAUSE, LAST MENSTRUAL PERIOD_____
BREAST LUMPS, LAST MAMMOGRAM_____

**FAMILY HISTORY** – HEART ATTACKS UNDER THE AGE OF 60, BLEEDING, BLOOD CLOTS, PROBLEMS WITH ANESTHESIA, OTHER_____

**SMOKING** (Y/N)   PPD:_____YEARS_____QUIT IN WHAT YEAR_____
**ALCOHOL:** NONE, MODERATE, HEAVY, IN RECOVERY, DT'S, SEIZURES
**SOCIAL HISTORY:** MARRIED, SEPARATED, DIVORCED, WIDOWED
EMPLOYED AS_____, RETIRED, HIGHEST LEVEL OF EDUCATION/DEGREES_____.

| Medications | Dose | Schedule |
|---|---|---|
|  |  |  |
|  |  |  |
|  |  |  |
|  |  |  |
|  |  |  |
|  |  |  |
|  |  |  |
|  |  |  |
|  |  |  |
|  |  |  |
|  |  |  |

| Prior Surgeries / Complications | Hospital | Date | Surgeon |
|---|---|---|---|
|  |  |  |  |
|  |  |  |  |
|  |  |  |  |
|  |  |  |  |
|  |  |  |  |
|  |  |  |  |
|  |  |  |  |
|  |  |  |  |
|  |  |  |  |

| ALLERGIES | TYPE OF REACTION |
|---|---|
|  |  |
|  |  |
|  |  |
|  |  |
| **Nuts/Foods-** |  |
| **Metal-** |  |
| **Latex-** |  |

Notes:_____

_____
_____
_____
_____
_____

# Table 11. Knee Society Score (questionnaire to be filled out by patient)

Name_____ Date of Birth_____
Today's Date_____ Date of Surgery_____

Circle the best answers:
**1- How much pain do you have when walking?**
- None
- Mild or occasional
- Moderate
- Severe

**2- How much pain does your knee cause when going up and down stairs?**
- None
- Mild or occasional
- Moderate
- Severe

**3- How much pain does your knee cause when you are at rest?**
- None
- Mild
- Moderate
- Severe

**4- How does your knee affect your walking ability?**
- I can walk unlimited distances
- I can walk 10-20 blocks
- I can walk 5-10 blocks
- I can walk 1-5 blocks
- I can walk less than one block
- I cannot walk at all

**5- How do you go up stairs?**
- I go up stairs normally with one foot in front of the other
- I use the hand rail for balance
- I use the hand rail to pull myself up
- I cannot climb stairs

**6- How do you go down stairs?**
- I go down stairs normally one foot in front of the other
- I use the hand rail for balance
- I use the hand rail to support myself
- I cannot go down stairs

**7- How do you get out of a chair?**
- I get out of a chair normally without support
- I use the arm rests for balance
- I use the arm rests to push myself
- I cannot get out of a chair

**8- What type of support do you use when walking?**
- None
- Cane
- 2 canes
- Crutches
- Walker

| Part I. Knee Score (KSS, Insall 1993) | points | score |
|---|---|---|
| **Pain (30 points)** | | |
| **Walking (Question #1)** | | |
| None | 35 | |
| Mild or occasional | 30 | |
| Moderate | 15 | |
| Severe | 0 | |
| **Stairs (Question #2)** | | |
| None | 15 | |
| Mild or occasional | 10 | |
| Moderate | 5 | |
| Severe | 0 | |
| **Range of motion (25 points maximum)** | | |
| 1 point for each 5° of arc of motion to a maximum of 25 points | Up to 25 | |
| **Stability** | | |
| **Medial/Lateral** | | |
| 0-5 mm | 15 | |
| 5-10mm | 10 | |
| >10mm | 5 | |
| **Anterior/Posterior** | | |
| 0-5mm | 10 | |
| 5-10mm | 8 | |
| >10mm | 5 | |
| More than 10° | | |
| **Deductions** | | |
| **Extension lag** | | |
| None | 0 | |
| ≤5 degrees | -2 | |
| 5-10 degrees | -5 | |
| >10 degrees | -10 | |
| **Flexion contracture** | | |
| ≤5 degrees | 0 | |
| 6-10 degrees | -3 | |
| 11-20 degrees | -5 | |
| >20 degrees | -10 | |
| **Malalignment** | | |
| -2 points for every 5 degrees (0 points if between 5-10°) | -2 points per 5 degrees | |
| **Pain at rest (Question #3)** | | |
| Mild | -5 | |
| Moderate | -10 | |
| Severe | -15 | |
| Symptomatic plus objective | 0 | |
| **Total Knee Score** | = | |

| Part II. Functional Score (KSS, Insall 1993) | points | score |
|---|---|---|
| **Walking (Question #4)** | | |
| Unlimited | 55 | |
| 10-20 blocks | 50 | |
| 5-10 blocks | 35 | |
| 1-5 blocks | 25 | |
| <1 block | 15 | |
| Cannot walk | 0 | |
| **Stairs up (Question #5)** | | |
| Normal | 15 | |
| Hands balance | 12 | |
| Hands pull | 5 | |
| Cannot or bizarre | 0 | |
| **Stairs down (Question #6)** | | |
| Normal | 15 | |
| Hands balance | 12 | |
| Hands hold | 5 | |
| Cannot or bizarre | 0 | |
| **Chair (Question #7)** | | |
| Normal | 15 | |
| Hands balance | 12 | |
| Hands push | 5 | |
| Cannot get out of chair | 0 | |
| **Deductions (Question #8)** | | |
| Cane | -2 | |
| Crutches | -10 | |
| Walker | -10 | |
| | | |
| **Functional Score (100 Max)** | = | |

# Table 12. Modified Harris Hip Score [134]

Name_____ Date of birth_____
Date of Surgery_____ Today's date_____

| | | |
|---|---|---|
| **I. Pain (44 points max)** | | |
| None or ignores it | 44 | |
| Slight, occasional, no compromise in activities | 40 | |
| Mild pain, no effect on average activities, rarely moderate pain with unusual activity, may take aspirin | 30 | |
| Moderate pain, tolerable but makes concessions to pain. Some limitation of ordinary activity or work. May require occasional pain medicine stronger than aspirin | 20 | |
| Marked pain, serious limitation of activities | 10 | |
| Totally disabled, crippled, pain in bed, bedridden | 0 | |
| **II. Function (47 points max)** | | |
|   **A. Gait (33 points max)** | | |
|     Limp | | |
|       None | 11 | |
|       Slight | 8 | |
|       Moderate | 5 | |
|       Severe | 0 | |
|     Support | | |
|       None | 11 | |
|       Cane for long walks | 7 | |
|       Cane most of the time | 5 | |
|       One crutch | 3 | |
|       Two canes | 2 | |
|       Two crutches | 0 | |
|       Not able to walk (specify reason) | 0 | |
|     Distance walked | | |
|       Unlimited | 11 | |
|       Six blocks | 8 | |
|       Two or three blocks | 5 | |
|       Indoors only | 2 | |
|       Bed and chair | 0 | |
|   **B. Activity (14 points max)** | | |
|     Stairs | | |
|       Normally without using a railing | 4 | |
|       Normally while using a railing | 2 | |
|       In any manner | 1 | |
|       Unable to do stairs | 0 | |
|     Shoes and socks | | |
|       With ease | 4 | |
|       With difficulty | 2 | |
|       Unable | 0 | |
|     Sitting | | |
|       Comfortably in ordinary chair one hour | 5 | |
|       On a high chair for one-half hour | 3 | |
|       Unable to sit comfortably in any chair | 0 | |
|     Enter public transportation | 1 | |
| **III. Absence of deformity** 4 points are given if the patient demonstrates less than 30° flexion contracture, less than 10° abduction contracture, less than 10° internal rotation contracture, limb length discrepancy less than 3.2 centimeters | 0 or 4 | |
| **IV. Range of motion** (modified for simplification, 5 points max) | | |
|   Flexion to 45 degrees (one point) | 1 | |
|   Flexion to 110 degrees (one additional point) | 1 | |
|   Abduction to 15 degrees | 1 | |
|   External rotation to 15 degrees | 1 | |
|   Adduction to 15 degrees | 1 | |

**Table 13. Tests to Evaluate Excessive Clotting (hypercoagulability)**

| Lab Test | Nature of Disease |
|---|---|
| Protein C | Hereditary |
| Protein S | Hereditary |
| Antithrombin 3 | Hereditary |
| Protein V Leiden | Hereditary |
| Prothrombin gene mutation (20210) | Hereditary |
| MTHFR | Hereditary |
| Lupus anticoagulant | Acquired |
| Antiphospholipid antibodies | Acquired |
| Beta$_2$ glycoprotein I Antibodies | Acquired |

**Table 14. Lab Abnormalities in Total Joint Patients with DVT/PE in Spite of DVT Prophylaxis** [112]

| Test | Patients with DVT | Controls (no DVT) | p value |
|---|---|---|---|
| Antithrombin III | 5/42 (12%) | 0/43 (0%) | 0.026 |
| Protein C | 9/42 (21%) | 2/43 (4.7%) | 0.021 |
| Prothrombin gene mutation (G20210) | 10/43 (23%) | 1/43 (2%) | 0.0037 |
| Factor V Leiden | 1/42 (2.4%) | 2/42 (4.8%) | ~1.0 (NS) |
| MTHFR | 12/43 (28%) | 20/42 (48%) | 0.075 (NS) |
| High homocysteine | 9/39 (23%) | 5/40 (15%) | 0.35 (NS) |

**Table 15. Quantified Risks for Infection in Joint Replacement in Selected Conditions**

| Risk Factor for Infection | Magnitude of Risk (references) |
|---|---|
| Rheumatoid Arthritis | 2.6 fold higher [120], 4% [121] |
| Obesity | 6.7 fold higher in total knee replacement [128] 4.2 fold higher in total hip replacement [128] |
| Diabetes | 0-7% in various studies (96-98) |
| Cirrhosis | 7% [29] |
| Renal failure / Dialysis | 13% [130] |
| AIDS | 14% [129] |
| IV Drug Use | 25% [129] |
| Prior Infection in Joint | 4% [126] |
| Prior Infection in Bone | 15% [126] |
| Prior Surgery on Joint | 2 fold higher for knee replacement [120] 3 fold higher for hip replacement [120] |
| Immunosuppression for Organ Transplantation | 19% [33] |
| Psoriasis | 17% [125] |

Table 16. DVT Prophylaxis in Total Hip Replacement: A Meta-Analysis of 10,929 Patients (Adapted form Freedman, et al.) [135]

| Agent | Distal (Calf) DVT | Proximal (Thigh) DVT | Total DVT | PE | Death (All Causes) | Major Bleeding (Wound Bleeding) |
|---|---|---|---|---|---|---|
| Low Molecular Weight Heparin | 9.6 | 7.7 | 17.7 | 0.36 | 0.20 | 2.22 (1.51) |
| Warfarin | 17.1 | 6.3 | 23.2 | 0.16 | 0.45 | 1.67 (0.76) |
| Aspirin | 19.7 | 11.4 | 30.6 | 1.28 | 0.29 | 0.73 (0.29) |
| Mechanical Compression | 7.7 | 13.3 | 20.7 | 0.26 | 1.01 | 0.00 (0.00) |
| No Prophylaxis | 22.4 | 25.8 | 48.5 | 1.51 | 0.33 | 0.56 (0.28) |

Table 17. DVT Prophylaxis in Total Knee Replacement: A Meta-Analysis of 3,482 Patients (Adapted for Brookenthal, et al.) [136]

| Agent | Distal (Calf) DVT | Proximal (Thigh) DVT | Total DVT | PE | Death (All Causes) | Major Bleeding (Wound Bleeding) |
|---|---|---|---|---|---|---|
| Low Molecular Weight Heparin | 24.4 | 5.9 | 31.3 | 0.2 | 0.2 | 2.4 |
| Warfarin | 35.6 | 10.2 | 45.6 | 0.4 | 0.6 | 1.3 |
| Aspirin | 55.2 | 1.7 | 56.9 | 0.0 | 0.0 | 0.0 |
| Mechanical Compression | 29.5 | 3.3 | 38.3 | 0.0 | 0.0 | 0.0 |
| No Prophylaxis | 60.2 | | | 0.0 | 0.0 | |

# Notes on DVT Prophylaxis

Hip replacement is associated with significantly higher rates of pulmonary embolism and death from its consequences than is knee replacement. Two of the agents, warfarin and aspirin, are inferior at preventing DVT in the calf and are often combined with a mechanical compression device, which has a superior track record at preventing distal (calf) DVT. Mechanical compression alone is inferior to medications in preventing proximal (thigh) DVT, and proximal DVT is believed to create a greater risk of pulmonary embolism and death. Low molecular weight heparin is superior at preventing DVT, pulmonary embolism, and death but is associated with lower blood counts, more bleeding, and higher transfusion rates [36]. The incidence of serious bleeding from the surgical wound with the use of low molecular weight heparin after total joint replacement is increased 50% over that seen in warfarin [137]. The American College of Chest Physicians believes that aspirin is inferior to the other agents and to the use of mechanical devices.

The actual incidence of pulmonary emboli in hip replacement has gone down over time and may not be as high as in the table above which includes data from as long ago as 1974. This is due to shorter operations, less blood loss, earlier mobilization (walking) and changes in anesthetic techniques. For example, the use of epidural anesthesia may prevent blood clots after total joint surgery [145]. Hypotensive anesthesia with drug support of the blood pressure with medications such as epinephrine increases the blood flow to the foot and may also be useful in preventing clots [138].

The duration of prophylaxis should be at least 4 weeks after hip replacement and at least two weeks after knee replacement. The average time to the discovery of a clot after hip replacement is 17 days, compared to 7 days for knee replacement [55].

# Table 18. Consumer Reports on Glucosamine and Chondroitin Amounts [139]

| Adequate (at least 90% of labeled amounts) | Glucosamine | Chondroitin | Cost per day |
|---|---|---|---|
| Kirkland Signature Extra Strength Glucosamine HCl and Chondroitin Sulfate (Costco) | 115% | 105% | $0.25 |
| Spring Valley Glucosamine & Chondroitin Double Strength (Wal-Mart) | 105 | 100 | 0.40 |
| Target Triple Strength Glucosamine & Chondroitin Complex | 110 | 95 | 0.45 |
| Vitamin World Glucosamine Chondroitin Double Strength | 105 | 100 | 0.45 |
| Vitasmart Double Strength Glucosamine & Chondroitin (Kmart) | 110 | 100 | 0.45 |
| Now Glucosamine & Chondroitin Extra Strength | 110 | 95 | 0.55 |
| GNC Glucosamine 750/Chondroitin 600 | 95 | 95 | 0.60 |
| Safeway Select Double Strength Glucosamine Chondroitin | 110 | 95 | 0.65 |
| CVS Glucosamine & Chondroitin Double Strength | 95 | 95 | 0.70 |
| Walgreens Finest Natural Glucosamine & Chondroitin Double Strength | 110 | 90 | 0.80 |
| Osteo Bi-Flex Glucosamine Chondroitin Triple Strength Complex | 95 | 100 | 0.90 |
| Nature's Bounty Extra Strength Glucosamine & Chondroitin Complex | 100 | 100 | 1.00 |
| CosaminDS Joint Health Supplement | 110 | 100 | 1.25 |
| **Marginally Adequate (80-90% of labeled amounts)** | | | |
| 21st Century Triple Strength Glucosamine & Chondroitin 3X | 80 | 80 | 0.60 |
| Glucoflex Glucosamine & Chondroitin Sulfate Triple Strength | 105 | 85 | 1.20 |
| **Inadequate (less than 80% of labeled amounts)** | | | |
| FlexAble Glucosamine & Chondroitin Sugar Free Chewables with Vitamin C | 110 | 60 | N/A |
| Trader Darwin's Glucosamine Chondroitin (Trader Joe's) | 110 | 10 | N/A |

### Table 19. The "Mini" Protocol for Anesthesia and Pain Management[144]

1. Intraoperative continuous bupicacaine plus fentanyl epidural.
2. Intraoperative sedation.
3. Intraoperative medications prior to incision: dexamethasone 12mg and ketorolac 15-30mg IV.
4. Epidural discontinued in operating room (administer 1-3mg duramorph prior to removal for knees only).
5. Oxycontin™ 10-20mg BID for 4 doses, first dose in recovery.
6. Celebrex 100-200mg daily, first dose in recovery.
7. Dexamethasone 10mg every 8 hours for 2 more doses.

# Glossary

Acetabulum – the part of the pelvis that forms the cup, or socket, of the hip joint (the femoral head forms the ball of this ball and socket joint).

Alternative bearing surfaces – the name given to new surface materials developed to create less wear than the original metal and plastic articulations. These include highly cross-linked polyethylene, metal-on-metal bearing and ceramic bearing surfaces.

Alumina – the ceramic derivative of the metal aluminum that provides a highly wear-resistant surface in some hip replacement systems. See also ceramic.

Arthritis – disease of the joint cartilage or the supporting tissues of joints. There are over 150 known types of arthritis, the most common form of which is osteoarthritis.

Arthropathy – a more specific name for types of arthritis not typically associated with inflammation. "Artho" means "joint," "-pathy" means "disease."

Arthroplasty – the surgical reconstruction of a joint, involving resurfacing or replacing of at least one surface in the joint.

Arthroscopy – the examination and repair of the interior of the knee joint with the use of an endoscope through a very small incision.

Articulation – the point where moving surfaces come together in an arthroplasty.

Augmentation Patella™ – a novel implant that replaces deficient kneecap bone. This device is made of the metal tantalum, the element used in Trabecular Metal™ implants.

Autoimmune disease – a disease in which one's immune system attacks one's own tissues. This disease is treated with medications that suppress the immune system.

Bedsores – ulcers to the skin that can be caused by lying in a bed for an extended period of time.

Bilateral – both sides.

Bone ingrowth – a crucial component for the success of cementless total joint implants, which require surrounding bone to grow into the pores of the implant. Some metals and designs used in implants are particularly receptive to bone ingrowth.

Bone sparing – surgical techniques that avoid disruption of the bone around a joint.

Bowlegged – see varus deformity.


C-reactive protein (CRP) – a circulating protein in the blood serum that increases with inflammation or infection.

Cardiovascular – referring to the blood vessels and heart.

Cartilage – the connective tissue that provides the surface for motion in healthy joint surfaces. Cartilage consists of cells and a matrix. The cells produce the matrix and reproduce to renew this tissue. The matrix contains large molecules that hold water and provide the cushioning of the joint. The matrix also includes collagen which maintains the integrity of the tissue and attaches cartilage tissue to its underlying bone.

Catheter – a flexible tube, such as the tube that is inserted into the urethra to collect urine.

Cementless fixation – the implantation of total joint hardware prostheses without the use of bone cement (polymethyl methacrylate, PMMA) to hold the implants in place. Cementless implants are incorporated into the bone through bone ingrowth.

Ceramic – a compound of metals and oxygen combined with trace amounts of other nonmetallic substances to improve their mechanical characteristics. Ceramics are highly wear-resistant bearing surfaces.

Components – any of the particular parts (implants) placed in the joint during an arthroplasty. Components are also referred to as implants or hardware, and the collective components and their placement is referred to as an articulation.

Contracture – a limit in one's range of motion, either in the amount of flexion or extension.

COX1 – the type of cyclooxygenase receptors (COX) found in blood platelets, the kidneys, the liver, and the GI tract. Both COX1 and COX-2 chemical receptors are blocked by certain non-selective anti-inflammatory medications used to treat arthritis. Some of the changes in the body's chemistry that occur from blocking these receptors are undesirable, such as ulcers and kidney disease. Cyclooxygenase is an enzyme that has normal roles in our bodies' chemistry as well as abnormal roles in inflammation in arthritis.

COX-2 – as above, these receptors are on muscles, joints, and tendons but also are present in the kidneys and probably in the cells that line blood vessels. Less likely to cause ulcers because they don't effect the GI tract, COX-2 inhibitors still can be associated with kidney disease and clotting problems that can lead to stroke and heart attack. Because the arthritis medications that are specific or selective blockers of COX-2 receptors are less likely to cause ulcers, these medications are still useful. Bleeding ulcers are the main cause of death from side effects of these medications.

Cross-linking – the use of radiation and heat to increase interconnections of polyethylene molecules to reduce certain types of wear to this plastic when it is used in hip replacement as a bearing surface. There may be benefit from cross-linking for knee replacement as well.

Delamination – flaking of the surface such as seen in knee replacement plastics, especially when they have been sterilized with radiation in air which creates oxidation of the surface of the plastic.

Debridement – the removal of foreign debris and dead tissue

from the surgical site. A debridement gets rid of matter that can potentially act as wear particles, and thus it is a counter-measure against osteolysis and synovitis.

Dementia – a loss of cognitive abilities such as may be seen in degenerative disorders such as Alzheimer's disease.

Diuretic – anything that causes one to pass fluid (urine) more frequently to eliminate excess water from the body.

Echocardiogram – a test that allows cardiologists to view the motion patterns of the muscles in the heart and heart valves.

Electrocardiogram – a test that records the electrical activity produced by the beating of the heart. This test demonstrates the rhythm of the heart as well as damage to the muscles such as can occur from a heart attack.

Endoscope – a small camera and its fiber optic lens such as the arthroscope which is used in arthroscopic knee surgery that allows the surgeon to see the anatomy of the knee from the inside, through small incisions.

Enzyme – a protein that acts like a catalyst in biochemical reactions.

Epidural – a method of giving anesthesia in which the anesthetic is injected into the fatty tissue that surrounds the nerve roots exiting the spine.

Erythrocyte sedimentation rate (ESR) – a measurement of inflammation or activity of infection based on the amount of proteins attached to the red blood cells.

Exhaust suits – suits that are generally worn by the surgeon performing a total joint replacement and the first assistant. These suits help contain possible contamination from the bodies of the surgical team working over the surgical site and thus lower the infection rate.

Fixation – the process of positioning and holding parts in the correct anatomical positions.

Fluoroscope – an x-ray machine that shows images in real

time, not requiring the x-ray films to be developed. This apparatus is helpful for injection of the hip joint.

Fracture – any of various types of bone breakage.

Fusion – fixation of the bones around a joint to convert them into one bone, which eliminates all motion in that joint.

Hemoglobin – the substance in red blood cells that carries oxygen throughout the body.

Hereditary – that which is transferred genetically from one generation to the next.

Hip liners – made of various materials (generally polyethylene, ceramic, or metal), the hip liner is placed inside the metal cup that fits into the acetabulum. These liners are made to be wear-resistant, and some have a special shape to prevent hip dislocation.

Implants – any device placed in the body such as the plastic, metal, and/or ceramic pieces that are used to resurface a joint surface in a total joint arthroplasty.

Ketoprofen gel – a non-steroidal anti-inflammatory gel that can be absorbed through the skin to a limited extent. This gel must be prepared by a pharmacist.

Knock-kneed – see valgus deformity.

Macrophage – scavenger cells that consume debris (wear particles) in the joint, causing the secretion of a hormone that stimulates bone resorptive cells (osteoclasts), which causes bone to dissolve in a process known as osteolysis.

Malalignment – failure to obtain ideal position of the bones or joints such as may occur with joint replacement. This can result in limping, dislocations, or premature wear.

Mechanical derangement – the loss of smooth surfaces, the development of loose bodies, or the loss of the integrity of a joint which can result in pain, locking, swelling, loss of alignment, or instability.

Mega prosthesis – a large prosthesis in which metal is used to replace whole segments of bone in the femur (available for both knee and hip reconstructions).

Mini-incision surgery – hip or knee replacement surgery through an incision of 4" or smaller.

Narcotic pain medications – controlled substances that are only available with a doctor's prescription. Narcotics are derivatives of opium, a material found in plants which mimic the body's pain control transmitters (endorphins).

Obesity (and morbid obesity) – the condition of being overweight. Obesity is diagnosed when the ratio of body mass and surface area of the skin (called the body mass index, or BMI) is over 40. When the BMI is over 50, morbid obesity is diagnosed. Morbid obesity is associated with an increased risk of death from medical problems that are associated with being severely overweight. BMI is calculated by dividing weight in kilograms by height in meters squared.
**$BMI = W(kg)/h(m)^2$.**

Osteoarthritis – the most common form of arthritis, which is associated with pain in the joints caused by the deterioration of cartilage on the joint surfaces. When the cartilage deteriorates significantly in a joint surface, the bones are exposed and rub against one another when that joint is put into motion. This causes pain and stiffness in the arthritic joint.

Osteoblast – bone cells that produce new bone. Abnormal osteoblast functions can result in fragile bone syndromes such as osteogenesis imperfecta, a disease unrelated to osteoporosis that causes fractures of bone from minimal trauma.

Osteoclast – bone cells that reabsorb old bone in normal bones but can work overtime in osteolysis and destroy bone. Over activity can also be stimulated by tumors and abnormal endocrine activity such as increases in thyroid or parathyroid hormones. Under activity of osteoclasts increases bone strength. In the extreme form of under activity, osteopetrosis (rock bone) occurs and bone marrow elements that make blood cells are deficient.

Osteolysis – dissolving bone caused (primarily) by wear particles that are absorbed by macrophages, which creates a hormone secretion that causes osteoclasts to dissolve the surrounding bone. Osteolysis occurs more commonly around hip implants than knee implants. Small wear particles and large numbers of wear particles are more dangerous in osteolysis.

Osteoporosis – a gradual loss of bone mass related to decreased hormone level in middle age, smoking, excess alcohol intake, and certain medications. Osteoporosis causes bones to become brittle and increases the likelihood of bone fractures.

Osteotomy – cutting through a bone. Used around the hip and knee to change the alignment of the joints.

Oxygen diffusion – the transport of oxygen from airways to blood vessels.

Physical derangement – see mechanical derangement.

Pigmentation – skin coloring. There may be increased coloring around an incision that is more visible for the first 1-2 years after surgery.

Platelets – a fragment of a blood cell (megakaryocyte) that circulates in the blood and initiates blood clotting when bleeding begins.

Polyethylene – a stable, highly wear-resistant plastic that is commonly used in total joint replacement implant systems.

Prednisone – a strong anti-inflammatory medicine that suppresses the immune system. This medication is a steroid, as apposed to non-steroidal anti-inflammatory medicines. This class of medications, when used for long periods of time, can lead to thin bones (osteoporosis), and elevate blood sugars, particularly in diabetic patients. Used in high doses for long periods of time, it can cause obesity in the chest, back, abdomen, and face, along with other physiologic changes collectively called Cushing's syndrome.

Prosthesis – any material that replaces or acts as a body part.

Resorption – the reuptake of a material such as bone, which occurs in osteolysis, for instance.

Rheumatoid arthritis – an immune system disease in which the cells that line the joints destroy the surrounding cartilage.

Scavenger cells – cells such as the macrophages that cause osteolysis.

Sciatica – spine pain referred into the buttock, thigh, back of the knee, calf, or foot from a spine condition.

Sepsis – systemic infection in which bacteria, fungus, or viruses spread through the blood stream. Urosepsis, for instance, spreads to the blood from a urinary tract infection.

Sleep apnea – episodic cessation of breathing during sleep. Obstructive sleep apnea (where the loss of muscle tone during REM sleep causes the tongue to fall into the airway) is more common than central sleep apnea (mediated through the nervous system). Often there are mixed components to sleep apnea.

Soft tissue calcification – formation of calcium deposits in the muscles or other soft tissues around a joint from severe inflammation. Heterotopic ossification is an example and is frequent after joint replacement, especially of the hip.

Surface ceramic – a technology used in some implants in which the wear surface is chemically converted to its ceramic derivative. For example, in a surface zirconia implant, the surface is zirconia (oxidized zirconium) while the rest of the material is zirconium, the elemental metal from which the ceramic was derived. Surface ceramics reduce the amount of wear particles as they are harder than the metals used in joint replacement. They are less prone to pitting and scratching than metals, and they have an ionized (charged) surface that attracts lubricating joint fluid.

Synovitis – inflammation of the lining of a joint as may occur in rheumatoid arthritis, when bleeding into the joint has occurred, from infection, or from wear particles, particularly the particles that are too big to be absorbed by macrophages.

Trabecular metal – a novel construction of the metal tantalum into a material used in joint replacement.

Tray dislocation – the dislocation of the polyethylene tray from the tibial component in knee replacement.

Unicompartmental knee replacement – a surgical procedure in which only one compartment (either the inside or the outside half) of the knee is replaced. Some surgeons prefer to use this surgery when the cartilage is still healthy on the other side of the knee.

Vascular – referring to the blood vessels and the circulation of blood.

Valgus deformity – a physical derangement in which the extremity is bent outward (forcing the joint toward the body's mid-line). In the knee, this is called a "knock-kneed" deformity.

Varus deformity – a physical derangement in which the extremity is bent inward (forcing the joint away from the body's mid-line). In the knee, this is called a "bow-legged" deformity.

Viscosupplementation – the injection of an artificial joint fluid made from rooster combs that can act as a reasonable non-surgical, short-term treatment for osteoarthritis.

Wear particle – any of various foreign particles—including bone, cement, metal, ceramic, or polyethylene—present in a joint. Large wear particles from materials such as polyethylene can cause synovitis. Smaller wear particles from metals, cement, and bone can cause osteolysis. The effects of very small wear particles from ceramics, for instance, are under study.

Zirconia – the ceramic derivative of the metal zirconium that is used as a highly wear-resistant surface ceramic in certain knee and hip implant systems. See also ceramic.

# References

1. Mahomed NN, Barrett JA, Katz J, Baron JA, Wright J, Losina E. Epidemiology of total knee replacement in the United States Medicare population. *J Bone Joint Surg Am* 2005;87:1222-28.
2. Mahomed NN, Barrett JA, Katz JN, Phillips CB, Losina E, Lew RA, Guadagnoli E, Harris WH, Poss R, Baron JA. Rates and outcomes of primary and revision total hip replacement in the United States Medicare population. *J Bone Joint Surg* 2003; 85:27-32.
3. National Vital Statistics Report. Vol. 53, No. 5. October 12, 2004.
4. Wolfe MM, Lichtenstein DR, Singh G. Gastrointestinal toxicity of nonsteroidal antiinflammatory drugs *NEJM*: 1888-1899, 1999 and Singh, Am J Med 1998:105(1B): S31-8
5. Lai K-A, Shen W-J, Yang C-Y, et al. The use of alendronate to prevent early collapse of the femoral head in patients with nontraumatic osteonecrosis. *J Bone Joint Surg Am.* 2005; 87: 2155-2159.
6. Evidence-based recommendations for the role of exercise in the management of osteoarthritis of the hip or knee - the MOVE consensus. Roddy E, Zhang W, Doherty M, et al. *Rheumatology* 2005; 44: 67-73.
7. Nevitt MC. Obesity outcomes in disease management: clinical outcomes for osteoarthritis. *Obesity Research* 2002; 10:32S-37S.
8. Heisel C, Silva M, Dela Rosa MA, Schmalzried TP. The effects of lower-extremity total joint replacement for arthritis on obesity. *Orthopedics* 2005; 28.2: 157-159.
9. Reginster JY, Deroisy R, Rovati LC, et al. Long-term effects of glucosamine sulphate on osteoarthritis

10. progression: a randomized, placebo-controlled clinical trial. *Lancet.* 2001: 357:251-256.
10. Clegg DO, Reda DJ, Harris CL, et al. The efficacy of glucosamine and chondroitin sulfate in patients with painful knee arthritis: the Glucosamine / chondroitin arthritis intervention trial (GAIT). 2005 ACR/ARHP Annual Scientific Meeting; November 12-17, 2005: San Diego CA. Presentation 622.
11. Abimbola OA, Cox DS Laing Z, et al. Analysis of glucosamine and chondroitin sulfate content in marketed products the caco-2 permeability of chondroitin sulfate raw materials. *JANA* 3:37-44.
12. Leopold SS, Redd BB, Warme WJ, et al. Corticosteroid compared with hyaluronic acid injections for the treatment of osteoarthritis of the knee. *J Bone Joint Surg Am.* 2003; 85:1197-1203.
13. Leopold SS, Warme WJ, Pettis PD, Shott S. Increased frequency of acute local reaction to intra-articular Hylan GF-20 (Synvisc) in patients receiving more than one course of treatment. *J Bone Joint Surg Am.* 2002; 84:1619-1623.
14. Forman JP, Stampfer MJ, Curhan GC. Non-narcotic alalgesic dose and risk of incident hypertension in US women. *Hypertension* 2005; 46:500-507.
15. Berman BM, Lao L, Langenberg P, Lee WL, Gilpin AMK, Hochberg MC. Effectiveness of acupuncture as adjunctive therapy in osteoarthritis of the knee: A randomized, controlled trial. *Annals of Internal Medicine* 2004; 141(12):901-910.
16. Felsh DT, Buckwalter J. Debridement and lavage for osteoarthritis of the knee. *NEJM* 2002; 347.2:132-133.
17. Blevins FT, Steadman JR, Rodrigo JJ, Silliman J. Treatment of articular cartilage defects in athletes: an analysis of functional outcome and lesion appearance. *Orthopedics* 1998; 21:761-68.

18. Muschler GF, Nitto H, Boehm CA, Easley KA. Age- and gender-related changes in the cellularity of human bone marrow and the prevalence of osteoblastic progenitors. *J Orthop Res* 2001;19(1):117-25.
19. Knutsen G, Engebretsen L, Ludvigsen TC, Drogset JO, Grøntvedt T, Solheim E, Strand T, Roberts S, Isaksen V, Johansen O. Autologous chondrocyte implantation compared with microfracture in the knee. A randomized trial. *J Bone Joint Surg Am* 2004; 86:455– 464.
20. Mithoefer K, Williams RJ, Warren RF, et al. The microfracture technique for the treatment of articular cartilage lesions in the knee: a prospective cohort study. *J Bone Joint Surg Am.* 2005; 87:1911-1920.
21. Yang, Niconson. Arthroscopic surgery of the knee in the geriatric patient. Clin Orthop 1995; 316: 50-58.
22. Morrey BF. Upper tibial osteotomy for secondary osteoarthritis of the knee. *J Bone Joint Surg Br* 1989; 71:554.
23. Mont MA, Dellon AL, Chen F, Hungerford MW, Krackow KA, and Hungerford DS. The operative treatment of peroneal nerve palsy. *J Bone Joint Surg Am* 1996; 78: 863-869.
24. Callaghan JJ, Brand RA, Pedersen DR. Hip Arthrodesis. *J Bone Joint Surg Am* 1985; 67:1328.
25. The effect of cane Use on hip contact force. Brand RA and Crowninshield RD. *Clin Orthop Rel Res* 1980; 147: 181-184.
26. *Consumer Reports*, "Joint replacement: 1001 patients tell you what your doctor can't," p15-18. June, 2006.
27. Lombardi AV, Mallory TH, Fada RA, et al. Simultaneous bilateral total knee arthroplasties, who decides? *Clin Orthop* 2001, 392: 319-329.
28. Lingard EA, Katz JN, Wright EA, Sledge CB. Predicting the outcome of total knee arthroplasty. *J Bone Joint Surg Am* 2004; 86:2179-2186.

29. Shih LY, Cheng CY, Chang CH, et al. Total knee arthroplasty in patients with liver cirrhosis. *J Bone Joint Surg Am* 2004; 86: 335-341.
30. Sakalkale DP, Hosack WJ, and Rothman RH. Total hip arthroplasty in patients on long-term renal dialysis. *J Arthroplasty* 1999; 14: 571-575.
31. van der Linden J, Lindvall G, Sartipy. Aprotinin decreases postoperative bleeding and number of transfusions in patients on clopidrogrel undergoing coronary artery bypass graft surgery. *Circulation* 2005; 112: I-276-I-280.
32. Rodgers A, Walker N, Schug S. Reduction of postoperative mortality and morbidity with epidural or spinal anaesthesia: Results from overview of randomised trials. *Brit Med J* 2000; 321: 1493-1497.
33. Tannenbaum DA, Matthews LS, Grady-Benson JC. Infection around joint replacements in patients who have a renal or liver transplantation. *J Bone Joint Surg Am* 1997; 79A:36-43.
34. Giles JT, Gelber AC, Nanda S, Barlett SJ, Bathon JM. TNF inhibitor therapy increase the risk of post-operative orthopedic infection in patients with rheumatoid arthritis. *Arthritis Rheum* 2004; 50(suppl 9) Abstract 1764.
35. Rosandich PA, Kelley JT, Conn DL. Perioperative management of patients with rheumatoid arthritis in the era of biologic response modifiers. *Current Opinion in Rheum* 2004; 16:192-98.
36. Bong MR, Patel V, Chang E, Issack PS, Hebert R, Di Cesare PE. Risks associated with blood transfusion after total knee arthroplasty. *J Arthroplasty* 2004; 10: 281-87.
37. Bierbaum BE, Callaghan JJ, Galante GO, Rubash HE, Tooms RE, and Welch RB. An Analysis of Blood Management in Patients Having a Total Hip or Knee Arthroplasty. *J Bone Joint Surg Am* 1999; 81:2-10.

38. Pierson JL, Hannon TJ, and Earles DR. A Blood-Conservation algorithm to reduce blood transfusions after total hip and knee arthroplasty. *J Bone Joint Surg Am* 2004;86:1512-18.
39. Cushner FD, Hawes T, Kessler D, et al. Orthopaedic-Induced Anemia, the fallacy of autologous donation programs. *Clin Orthop* 2005; 431: 145-149.
40. Bexwada HP, Nazarian DG, Henry DH, and Booth RE. Preoperative use of recombinant human erythropoietin before total joint arthroplasty. *J Bone Joint Surg Am* 2003; 85: 1795-1800.
41. *Watchtower.* 121: 12, p.29-31. 2000.
42. Parker MJ, Roberts CP, Hay D. Closed Suction Drainage for Hip and Knee Arthroplasty. A Meta-Analysis. *J Bone Joint Surg Am* 2004; 86:1146-1152.
43. Goodnough LT, Despotis GL, Merkel K, and Monk TG. A randomized trial comparing acute normovolemic hemodilution and preoperative autologous blood donation in total hip arthroplasty. *Transfusion* 2000; 40(9): 1054-1057.
44. Lee G-C, Cushner FD. Blood Management in Patients with Deep Prosthetic Hip and Knee Infections. *Orthopedics* 2004 27:1-5.
45. http://www.cancer.org/docroot/ETO_1_4x_Possbile_Risks_of_Blood_Product_Transfusions.asp?sitearea=ETO (Revised 11/28/06)
46. Adapted from John J. Callaghan, MD, "Blood Management in THR," Current Concepts in Joint Replacement Spring, 2000, Las Vegas, NV
47. http://www.nsc.org/lrs/statinfo/odds.htm
48. Kendall, SJH, Weir J, Aspinall R, et al. Erythrocye transfusion caused immunosuppression after total hip replacement. *Clin Orthop* 2000; 381: 145-155.
49. Hatzidakis AM, Mendlick MR, McKillip T, Reddy RL, and Garvin KL. Preoperative Autologous Donation for Total Joint Arthroplasty. An Analysis of

Risk Factors for Allogenic Transfusion. *J Bone Joint Surg Am* 2000; 82:89-100.
50. Neut D, van de Belt, Stokroos, van Horn, van der Meir, Busscher, Biomaterial-associated infection of gentamycin-loaded PMMA beads in orthopedic revision surgery. *J Antimicrob Chemo* 47: 885-891
51. Sliva CD, Callaghan JJ, Goetz DD, and Taylor SG. Staggered bilateral total knee arthroplasty performed four to seven days apart during a single hospitalization. *J Bone Joint Surg Am* 2005; 87: 508-513.
52. Keating EM, Meding JB, Faris PM, Ritter MA. Predictors of transfusion risk in elective knee surgery. *Clin Orthop* 1998; 357: 50-59.
53. Wolff LH, Parvizi J, Trousdale RT, Pagnano MW, Osmon DR, Hanssen AD, Haidukewych GJ. Results of treatment of infection in both knees after bilateral total knee arthroplasty. *J Bone Joint Surg Am* 2003; 8 5A:1952-55.
54. Hanssen AD, Rand JA. Evaluation and treatment of infection at the site of a total hip or knee arthroplasty. Instructional course lecture, American Academy of Orthopedic Surgeons. *J Bone Joint Surg Am* 1998; 80A:910-922.
55. White RH, Romano PS, Zhou H, Rodrigo J, Bargar W. Incidence and time course of thromboembolic outcomes following total hip or knee arthroplasty. *Arch Intern Med* 1998; 158: 1525-1531.
56. Eikenboom JW, Quinlan DJ and Douketis JD. Extended-duration prophylaxis against venous thrombolembolism after total hip or knee replacement: a meta-analysis of the randomised trials. *Lancet* 2001; 358: 9-15.
57. Lieberman JR and Hsu WK. Current Concepts Review. Prevention of venous thromboembolic disease after tot hip and knee arthroplasty. *J Bone Joint Surg*

*Am* 2005; 87: 2097-2112.
58. Parvizi J, Sullivan TA, Trousdale RT, and Lewallen DG. Thirty-day mortality after total knee arthroplasty. *J Bone Joint Surg Am* 2001; 83: 1157-1161.
59. Asp JP, and Rand JA. Peroneal nerve palsy after total knee arthroplasty. *Clin Orthop* 1990; 261: 233-237.
60. Nelson JD, Urban JA, Salsbury TL, et al. Acute colonic pseudo-obstruction (Ogilvie Syndrome) after arthroplasty in the lower extremity. *J Bone Joint Surg Am* 2006; 88: 604-610.
61. GAO Report to the Senate Ways and Means Committee: More Specific Criteria Required to Classify Inpatient Rehabilitation Facilities, GAO 05-366, April, 2005.
62. Mont MA, Rajadhyaksha AD, Marxen JL, Silbertein CE, Hungerford DS. Tennis after total knee arthroplasty. *Am J Sports Med* 2002; 30:163-166.
63. Rand JA, Trousdale RT, Ilstrup DM, Harmsen WS. Factors affecting the durability of primary total knee prosthesis. *J Bone Joint Surg Am* 2003; 85:259-263.
64. Insall JN, Ranawat CS, Aglietti P, Shine J. A comparison of four models of total Knee-replacement prostheses. *J Bone Joint Surg Am* 1976, 58: 754-765.
65. Ridgeway S, Moskal JT. Early instability with mobile-bearing total knee arthroplasty. A series of 25 cases. *J Arthroplasty* 2004; 19(6): 686-693.
66. Bert JM. Dislocation/subluxation of meniscal bearing elements after New Jersey Low-Contact Stress total knee arthroplasty. *Clin Orthop* 1990; 254:211.
67. Callaghan JJ, Insall JN, Greenwald AS, et al. Mobile-bearing knee replacement. Concepts and results. *J Bone Joint Surg Am* 2000; 82A:1020-1041.
68. Callaghan JJ, Squire MW, Goetz DD, et al. Cemented rotating-platform total knee replacement. A nine to twelve-year follow-up study. *J Bone Joint Surg Am* 82:705-711.

69. Callaghan JJ, O'Rourke MR, Iossi MF, et al. Cemented rotating-platform total knee replacement. A concise follow-up, at a minimum of fifteen years, of a previous report. *J Bone Joint Surg Am* 2005; 87:1995-1998.
70. Huang CH, Ma HM, Liau JJ, et al. Osteolysis in failed total knee arthroplasty: a comparison of mobile-bearing and fixed-bearing knees. *J Bone Joint Surg Am* 2002; 84A:2224-2229.
71. Fisher DA, Watts M, and Davis KE. Implant position in knee surgery: A comparison of minimally invasive, open unicompartmental and total knee arthroplasty. *J Arthroplasty* 2003; 18 (7 suppl 1): 2-8.
72. Heck DA, Marmor L, Gibson A, Rougraff BT. Unicompartmental knee arthroplasty. A multicenter investigation with long-term follow-up evaluation. *Clin Orthop* 1993; 286: 154-159.
73. Laurencin CT, Zelicof SB, Soctt RD, Ewald FC. Unicompartmental versus total knee replacement in the same patient. A comparative study. *Clin Orthop* 1991; 273: 151-156.
74. Hallock RH, Fell BM. Unicompartmental tibial hemiarthroplasty: Early results of the UniSpacer knee. *Clin Orthop* 2003; 416: 154-163.
75. Woolson ST, Mow CS, Syquia JF, Lannin JV. Comparison of primary total hip replacements performed with a standard incision or a mini-incision. *J Bone Joint Surg Am* 2004; 86:1353-58.
76. Dalury DF and Dennis DA. Mini-incision total knee arthroplasty can increas risk of component malalignment. *Clin Orthop* 2005; 440: 77-81.
77. Laskin RS. Minimally invasive total knee arthroplasty: the results justify its use. *Clin Orthop* 2005; 440: 54-59.
78. Kolisek FR, Bonutti PM, Hozack WJ, et al. Clinical experience using a minimally invasive surgical

approach for total knee arthroplasy: Early results of a prospective randomized study compared to a standard approach. *Arthroplasty* 2007; 22: 8-13.
79. Reis MD, Cabalo A, Bozic KJ, Anderson M. Porous tantalum patellar augmentation: The importance of residual bone stock. *Clin Orthop* 2006; 452: 166-170.
80. Paletta GA and Laskin RS. Total knee arthroplasty after a previous patellectomy. *J Bone Joint Surg Am* 1995; 77: 1708-1712.
81. Laskin RS, O'Flynn HM. Total knee replacement with posterior cruciate retention in rheumatoid arthritis: problems and complications. *Clin Orthop* 1997; 345: 24-28.
82. Babis GC, Trousdale RT, Morrey BF. The effectiveness of isolated tibial insert exchange in revision total knee arthroplasty. *J Bone Joint Surg Am* 2002; 84: 64-68.
83. Katz JN, Barrett J, Mahomed NN, Baron JA, Wright RJ, and Losina E. Association between hospital and surgeon procedure volume and the outcomes of total knee replacement. *J Bone Joint Surg Am* 2004; 86:1909-1916.
84. Ranawat CS, Rodriguez JA. Functional leg-length inequality following total hip arthroplasty. *J Arthroplasty* 1997; 12: 359-364.
85. Weeden SH, Paprosky WG, Bowling JW. The early dislocation rate in primary total hip arthroplasty following the posterior approach with posterior soft-tissue repair. *J Arthroplasty* 2003; 18: 709-713.
86. Westrich GH, Salvati EA, Sharrock N, Potter HG, Sanchez PM, Sculco TP. The effect of intraoperative heparin administered during total hip arthroplasty on the incidence of proximal deep vein thrombosis assisted by magnetic resonance venography. *J Arthroplasty* 2005; 20:42-50.
87. Westrich GH, Salvati EA, Sharrock N, et al. The effect of intraoperative heparin administered during

total hip arthroplasty on the incidence of proximal deep vein thrombosis assessed by magnetic resonance venography. *J Arthroplasty* 2005 20:42-50.
88. Barrett J, Losine E, Baron JA, et al. Survival following total hip replacement. *J Bone Joint Surg Am* 2005; 87:1965-1971.
89. Incavo SJ, Havener T, Benson E, et al. Efforts to improve cementless femoral stems in THR, 2- to 5-year follow-up of a high-offset femoral stem with distal stem modification (Secur-Fit Plus). *J Arthroplasty* 2004;19: 61.
90. Farrell CM, Springer BD, Haidukewych, GJ, Morrey BF. Motor nerve palsy following primary total hip arthroplasty. *J Bone Joint Surg Am* 2005; 87: 2619-2625.
91. Peak EL, Parvizi J, Ciminello M, Purtill JJ, Sharkey PF, Hozack WJ, Rothman RH. The role of patient restrictions in reducing the prevalence of early dislocation following total hip arthroplasty. *J Bone Joint Surg Am* 2005; 87:247-253.
92. Allain J, Roudot-Thoraval F, Delecrin J, et al. Revision total hip arthroplasty performed after fracture of a ceramic femoral head. *J Bone Joint Surg Am* 2003; 85:825-830.
93. Davies AP, Willert HG, Campbell PA, et al. An unusual lymphocytic perivascular infiltration in tissues around contemporary metal-on-metal joint replacements. *J Bone Joint Surg Am* 2005; 87: 18-27.
94. Jacobs JJ, Hallah NJ, Loosening and osteolysis associated with metal-on-metal bearings: a local effect of metal hypersensitivity? (Ed). *J Bone Joint Surg Am* 2006; 88: 1171-1172.
95. Milosev I, Trebse R, Koval S, et al. Survivorship and retrieval analysis of Sinkomet metal-on-metal total hip replacements at a mean of seven years. *J Bone Joint Surg Am* 2006; 88: 1173-1182.

96. Korovessis P, Petsinis G, Repanti M and Repantis T. Metallosis after contemporary metal-on-metal total hip arthroplasty. *J Bone Joint Surg Am* 2006; 88: 1183-1191.
97. Hallab NJ, Anderson S, Caicedo M, et al. Immune responses correlate with serum-metal in metal-on-metal hip arthroplasty. *J Arthroplasty* 2004; 19 (8 Suppl 3): 88-93.
98. Ladon D, Doherty A, Newson R, et al. Changes in metal levels and chromosome aberrations in the peripheral blood of patients after metal-on-metal hip arthroplasty. *J Arthroplasty* 2004; 19 (8 Suppl 3): 78-83.
99. Tharani R, Dorey FJ, and Schmalzried TP. The risk of cancer following total hip or knee arthroplasty. *J Bone Joint Surg Am* 2001; 83:774-780.
100. Beaule PE, LeDuff, M, Campbell P, Dorey FJ, Park SH, Amstutz HC. Metal-on metal surface arthroplasty with a cemented femoral component: A 7-10 year follow-up study. *J Arthroplasty* 2004; 19:17-22.
101. Nishii T, Nobuhiko S, Hidenobu M, et al. Does Alendronate Prevent Collapse in Osteonecrosis of the Femoral Head? *Clin Orthop* 2006; 443: 273-279.
102. Managing the care of patients with bisphosphonate-associated osteonecrosis: an American Academy of Oral Medicine position paper. Migliorati CA, Casiglia J, Epstein J, et al. *JADA* 2005; 136: 1658-1668
103. Adili A and Trousdale RT. Femoral head resurfacing for the treatment of osteonecrosis in the young patient. *Clin Orthop* 2003; 417: 93-101.
104. Orgonda L, Wilson R, Archibald P, Lawler M, Humphreys P, O'Brien, S, Nederland D. A minimal-incision technique in total hip arthroplasty does not improve early postoperative outcomes. *J Bone Joint Surg Am* 2005 87:701-710.
105. Mow CS, Woolson ST, Ngarmukos SG, et al.

Comparison of scars from total hip replacements done with a standard or mini-incision. *Clin Orthop* 2005; 441: 80-85.
106. Pagnano MW, Leone J, Lewallen DG, Hanssen AD. Two-incision THA had modest outcomes and some substantial complications. *Clin Orthop* 2005; 441: 86-90.
107. Berger RA. Total hip arthroplasty using the minimally invasive two-incision approach. *Clin Orthop* 417: 232-241, 2003.
108. Mardones R, Pagnano MW, Nemanich JP, and Trousdale RT. Muscle damage after total hip arthroplasty done with the two-incision and mini-posterior techniques. *Clin Orthop* 2005; 441: 63-67.
109. Katz JN, Losina E, Barrett J, Phillips CB, Mahomed NN, Lew RA, Guadagnoli E, Harris WH, Poss R, and Baron JA. Association between hospital and surgeon procedure volume and outcomes of total hip replacement in the United States Medicare population. *J Bone Joint Surg Am* 2001; 83:1622-1629.
110. Matta JM, Shahrdar C, Ferguson T. Single-incision anterior approach for total hip arthroplasty on an orthopedic table. Clin Orthop 2005; 441: 115-124.
111. Pellegrini VD, Clement D, Lush-Ehman C, et al. The natural history of thromboembolic disease after total joint arthroplasty. The case for routine surveillance as a contemporary management strategy. *Orthop Trans* 1994: 18:124.
112. Salvati EA, Gonzalez Della Valle, Westrich GH, et al. Heriatable thrombophilia and development of thromboembolic disease after total hip arthroplasty. *Clin Orthop* 2005: 41-55.
113. Incidence of undiagnosed sleep apnea in patients scheduled for elective total joint arthroplasty. Harrison MM, Childs A, Carson PE. *J Arthroplasty*

2003 17:1044-1047.
114. Parikh SN, Stuchin SA, Maca C, et al. Sleep apnea syndrome in patients undergoing toaal joint arthroplasty. *J Arthroplasty* 2002; 17
115. Edwards BN, Tulles HS, and Noble PC. Contributory factors and etiology of sciatic nerve palsy in total hip arthroplasty. *Clin Orthop* 1987; 218: 136-141.
116. Springer BD, Berry DJ, Lewallen DG. Treatment of periprosthetic femoral fractures following total hip arthroplasty with femoral component revision. *J Bone Joint Surg Am* 2003, 85:2156-2162.
117. Bhandari M, Guyatt GH, Griffith, Busse JW, Schünemann H. Effect of bisphosphonates on periprosthetic bone mineral density after total joint arthroplasty. A meta-analysis. *J Bone Joint Surg Am* 2005, 87:293-301.
118. Woolson ST, Northrup GD. Mobile- versus fixed-bearing total knee arthroplasty. *J Arthroplasty* 2004 19:135-140.
119. Kim YH, Kook HK, Kin JS. Comparison of fixed-bearing and mobile-bearing total knee arthroplasties. *Clin Orthop* 2001; 392:101-115.
120. Poss R, Thornhill TS, Ewald FC, et al. Factors influencing the incidence and outcome of infection following total joint arthroplasty. *Clin. Orthop* 1984; 182: 117-126.
120. Perl, TM, Cullen JJ, Wenzel RP, et al. Intranasal mucpirocin to prevent postop staphylococcus aureus infections. *NEJM* 2002; 346: 1871-1877. (see accompanying editorial)
121. Wilson MG, Kelley K and Thornhill TS. Infection as a complication of total knee-replacement arthroplasty. Risk factors and treatment in sixty-seven cases. *J. Bone Joint Surg Br* 1990; 72: 878-893.
122. England SP, Stern SH, Insall JN, and Windsor RE.

Total knee arthroplasty in diabetes mellitus. *Clin Orthop* 1990; 260: 130-134.
123. Papegelopoulos PJ, Idusuyi, Wallrichs SL, and Morrey BF. Long term outcome and survirorship analysis of primarty total knee arthroplasty in patients with diabetes mellitus. *Clin Orthop* 1996; 330: 124-132.
124. Menon TJ, Thjellsen D, and Wroblewski BM. Charnley low-friction arthroplasty in diabetic patients. *J Bone Joint Surg Br* 1983; 65: 580-581.
125. Stern SH, Insall JN, Windsor RH, et al. Total knee arthroplasty in patients with psoriasis. *Clin Orthop* 1989; 248: 108-111.
126. Luessenhop CP, Higgins LD, Brause BD and Ranawat CS. Multiple prosthetic infections after total joint arthroplasty. Risk factor analysis. *J Arthroplasty* 1996; 11: 862-868.
127. Surin VV, Sundholm K and Backman L. Infection after hip replacement. With special reference to discharge from the wound. *J Bone Joint Surg Br* 1983; 65: 412-418.
128. Namba RS, Paxton L, Fithian DC, and Stone ML. Obesity and Perioperative morbidity in total hip and total knee arthroplasty patients. J *Arthroplasty* 2005, 20 (7 Suppl 3): 46-50.
129. Lehman CR, Ries MD, Paiement GD, and Davidson AB. Infection after total joint arthroplasty in patients with human immunodeficiency virus or intravenous drug use. *J Arthroplasty* 2001; 16: 330-335.
130. Di Cesare PE, Chang E, Preston CF, and Liu C. Serum Interleukin-6 as a Marker of Periprosthetic Infection Following Total Hip and Knee Arthroplasty. *J Bone Joint Surg Am* 2005; 87:1921-1927.
131. Johnson DH, Fairclough JA, Brown EM and Morris R. Rate of bacterial recolonization of the skin after preparation: four methods compared. *British J Surg* 1987; 74: 64.

133. Rieker CB, Schon R, and Kottig P. Development and validation of a second-generation metal-on-metal bearing: laboratory studies and analysis of retrievals. *J Arthroplasty* 2004; 19 (8 Suppl 3): 41-44.
134. Harris WH. Traumatic arthritis of the hip after dislocation and acetabular fractures: treatment by mold arthroplasty. An end-result study using a new method of result evaluation. *J Bone Joint Surg Am* 1969; 51:737-55.
135. Freedman KB, Brookenthal KR, Fitzgerald RH, et al. A meta-analysis of thromboembolic prophylaxis following elective total hip arthroplasty. *J Bone Joint Surg Am* 2000; 82: 929-937.
136. Brookenthal KR, Freedman KB, Lotke PA, et al. A meta-analysis of thromboembolic prophylaxis in total knee arthroplasty. *J Arthroplasty* 2001; 16: 293-300.
137. Spiro TE, Johnson GJ, Christie MJ, et al. Efficacy and safety of enoxaparin to prevent deep venous thrombosis after hip replacement surgery. Enoxaparin Clinical Trial Group. *Ann Intern Med* 1994: 121: 81-89.
138. Sculco TP, Colwell CW, Pellegrini VD, et al. Prophylaxis against venous thromboembolic disease in patients having a total hip or knee arthroplasty. *J Bone Joint Surg Am* 2002; 84: 466-476.
139. *Consumer Reports*, "Joint supplements: brands to try and brands to avoid," p19-20. June, 2006.
140. Von Knoch M, Berry DJ, Harmsen WS, Morrey BF. Late dislocation after total hip arthroplasty. *J Bone Joint Surg Am* 2002; 84:1949-1953.
141. Stewart PS, Carlson RP. COLL 021 "Structure, interations and reactivity at microbial surfaces," presented at American Chemical Society 232nd national meeting, San Francisco, Sep 10, 2006
142. Singh AK, Szczech L, Tang KL, et al. Correction of anemia with epoetin alfa in chronic kidney disease.

*NEJM* 2006; 355: 2085-2098.
143. Ben-Galim P, Ben-Galim T, Rand N, et al. Hip-spine syndrome: The effect of total hip replacement surgery upon low back pain. #474. Presented at the American Academy of Orthpaedic Surgeons 74th Annual Meeting. Feb 14-18, 2007. San Diego.
144. Nuelle DG, Mann K. Minimal incision protocols for anesthesia, pain management, and physical thapy with standard incision in hip and knee arthroplasties: The effect on early outcomes. *Arthroplasty* 2007; 22: 20-25.
145. Sohn RS, Micheli LJ. The effect of running on the pathogenesis of osteoarthrits of the hips and knees. *Clin Orthop* 1985; 198:106-109.
146. Parvizi J, Viscusi ER, Frank HG, et al. Can Epidural anesthesia and warfarin be coadministered? Clin Ortho 2007; 456: 133-137.

# Trademarks

Actonel™. Aventis Pharmaceuticals, Inc. Bridgewater, NJ, USA.
Aleve™. Bayer, Inc. Etobicoke, Ontario, Canada.
Amethopterin™. Sigma-Aldrich Corp. St. Louis, MO, USA.
Arixtra™. GlaxoSmithKline, Inc. Brentford, Middlesex, UK.
Arava™. Aventis Pharmaceuticals, Inc. Bridgewater, NJ, USA.
Azasan™. Salix Pharmaceuticals, Inc. Morrisville, NC, USA.
Azulfidine™. Pharmacia & Upjohn, Inc. Kalamazoo, MI, USA.
Arthrotec™. Pfizer, Inc. New York, NY, USA.
Augmentation Patella™. Zimmer, Inc. Warsaw, IN, USA.
Bextra™. Pfizer, Inc. New York, NY, USA.
Celebrex™. Pfizer, Inc. New York, NY, USA.
Coumadin™. Bristol-Myers Squibb Company. Princeton, NJ, USA.
Darvocet™. aaiPharma, Inc. Wilmington, NC, USA.
Demerol™. Sanofi Winthrop Pharmaceuticals. New York, NY, USA.
Dilaudid™. Abbott Laboratories. Abbott Park, IL, USA.
Disalcid™. 3M Pharmaceuticals. St. Paul, MN, USA.
Enbrel™. Immunex Corp. Seattle, WA, USA.
EPO™. Amgen, Inc. Thousand Oaks, CA, USA.
Epoch™. Zimmer, Inc. Warsaw, IN, USA.
Epogen™. Amgen, Inc. Denver, CO, USA.
Evista™. Eli Lilly and Company. Indianapolis, IN, USA.
Folex™. Abbott Laboratories. Abbott, IL, USA.
Folex PFS™. Abbott Laboratories. Abbott, IL, USA.
Fosamax™. Merck & Co., Inc. Whitehouse Station, NJ, USA.
Fragmin™. Pharmacia & Upjohn, Inc. Kalamazoo, MI, USA.
Humira™. Abbott Laboratories. Abbott Park, IL, USA
Imperion™. Smith & Nephew, Inc. Memphis, TN, USA.
Imuran™. GlaxoWellcome. Auckland, New Zealand.
Innohep™. Leo Laboratories Canada Ltd. Ajax, Ontario, Canada.
Kineret™. Amgen, Inc. Denver, CO, USA.
Lodine™. Apotex, Inc. Weston, Ontario, Canada.
Lortab™. UCB Pharmaceuticals, Inc. Atlanta, GA, USA.
Lovenox™. Aventis Pharmaceuticals, Inc. Bridgewater, NJ, USA.
Mobic™. Boehringer Ingelheim Pharmaceuticals, Inc. Ridgefield, CT, USA.
Oxinium™. Smith & Nephew, Inc. Memphis, TN, USA.
Oxycontin™. Purdue Pharma L.P. Stamford, CT, USA.
Percocet™. Endo Pharmaceutical. Chadds Ford, PA, USA.
Plaquenil™. Sanofi Winthrop Pharmaceuticals. New York, NY, USA.
Plavix™. Sanofi Winthrop Pharmaceuticals. New York, NY, USA.

Procrit™. Ortho Biotech Products L.P. New Brunswick, NJ, USA.
Relafen™. GlaxoSmithKlein, Inc. Brentford, Middlesex, UK.
Remicade™. Centocor Pharmaceuticals, Inc. Horsham, PA, USA.
Rituxan™. Genentech, Inc. & Idec Pharmaceuticals, Inc. San Francisco, CA, USA.
Rheumatrex™. Pfizer, Inc. New York, NY, USA.
Silvidene™. Monarch Pharmaceuticals. Bristol, TN, USA.
Steri-Strips™. 3M Corp. St. Paul, MN, USA.
Sulzer™. Sulzer Orthopedics. Renamed Centerpulse Orthopedics. Austin, TX, USA.
Tamoxifen™. Apotex, Inc. Weston, Ontario, Canada.
Trabecular Metal™. Zimmer, Inc. Warsaw, IN, USA.
Trexall™. Barr Laboratories, Inc. Forest, VA, USA.
Tylenol™. McNeil-PPC. Fort Washington, PA, USA.
Tylox™. Ortho-McNeil Pharmaceutical, Inc. Raritan, NJ, USA.
Ultracet™. Ortho-McNeil Pharmaceutical, Inc. Raritan, NJ, USA.
Ultram™. Ortho-McNeil Pharmaceutical, Inc. Raritan, NJ, USA.
Vicodin™. Knoll Pharmaceutical. Mt. Olive, NJ, USA.
Vioxx™. Merck & Co., Inc. Whitehouse Station, NJ, USA.
Zyvox™. Pharmacia & Upjohn Inc. Kalamazoo, MI, USA.
Zyrconia™. Zimmer, Inc. Warsaw, IN, USA.

# Index

## A

acupuncture *20*
airport security *123*
all-cemented knee *75*
all-ceramic hip *97, 155*
all-metal hips *98*
all-polyethylene tibial component *76*
alternative bearing surfaces *70, 153, 175*
anemia *109, 111*, 128
anorexia *111*
anterior-lateral approach *106*
anti-inflammatory medications *4, 14, 18, 34, 79*
antibiotics *144, 146, 147*
arthropathy *1, 5, 6*
arthroscopic surgery of the knee *20*
aspirin *15, 16, 34, 91*
augmentation Patella *74*

## B

blood pressure *111*
blood transfusion *43, 109*
braces *19*
breathing problems *112*

## C

C-reactive protein (CRP) *147, 176*
cardiac problems *111*
cartilage *1, 9*
celebrex™ *14*
cementless fixation *79, 86, 176*
ceramic *124, 153-54, 158, 176*
chest pain *114*
chondroitin Sulfate *9*
computer navigated surgery *52, 74*
confusion *36, 61*
constipation *112*

contractures *52, 75*
cortisone *11*
COX-2 selective *14, 16*

## D

debridement *145, 177*
deep venous thrombosis *109*
delamination *74, 152, 177*
dementia *83, 178*
dental work *57, 123*
depression *117*
diarrhea *112*

## E

echocardiogram *32, 178*
electrocardiogram *31, 178*
erythrocyte sedimentation rate *125, 148*
exhaust suits *56, 85, 178*

## F

flexible composite hip *102*
fusion surgery *25*

## G

GI bleeding *15, 116*
glucosamine *2, 9*

## H

Harris Hip Score *168*
high volume surgeons and hospitals *76*

## I

immunosuppressive medications *40, 143, 160*
ingrowth femoral stems *95*
iron *109*

## K

ketoprofen gel *18, 179*
Knee Society Score *165*

# L

lateral approach  *106*

# M

macrophage  *153, 179*
malalignment  *3, 28, 179*
mechanical derangement  *20, 179=*
mega prosthesis  *131, 140, 180*
metal-on-metal  *87, 124, 158*
metal allergies  *59, 87, 99, 124*
microfracture technique  *22*
mini-incision surgery  *79, 103, 180*
mobile bearing and rotating platform knees  *69, 70*

# N

narcotic pain medications  *18, 117, 180*
neurological disorders  *116*
NSAIDs  *14, 160*

# O

obesity  *57, 83, 143, 180*
osteoarthritis  *1, 5, 7*
osteoblast  *180*
osteoclast  *153, 180*
osteolysis  *152, 153*
osteoporosis  *1, 137*
osteotomy  *24, 132, 181*
over-the-counter medications  *2, 9, 65*
oxinium™  *68, 97, 154, 158*
oxygen saturation monitor  *113*

# P

past medical history form  *162*
patellofemoral resurfacing  *71*
periprosthetic fractures  *137*
pigmentation  *58, 181*
polished tapered cemented stems  *96*
polyethylene  *95, 96, 125, 152, 181*
posterior-lateral approach  *106*
prednisone  *40, 41, 181*

pulmonary embolism  *57, 117*

# R

resorption  *138, 152, 182*
resurfacing arthroplasty  *100*
rheumatoid arthritis  *2, 5, 182*
risks of Blood Transfusion  *161*
risks of hip replacement  *81*
risks of knee replacement  *55*
rotating platform knees  *69*

# S

shortness of breath  *114*
small incision surgery and quadriceps sparing technique  *72*
surface ceramics (oxidized zirconium)  *68*

# T

tantalum  *103, 157*
total knee replacement without patellar (kneecap) resurfacing  *70*
Trabecular Metal™  *103, 157, 175*
tray dislocation  *69, 183*
trochanteric osteotomy  *106*
Tylenol™  *12*

# U

ulcers  *13, 15, 90, 116*
unicompartmental knee replacement  *71*
urological surgery  *57*

# V

valgus (knock-knee) deformity  *24, 51, 60, 183*
varus (bow-legged) deformity  *24, 142, 183*
Vioxx™  *14*
viscosupplementation  *10*
vtamins  *90, 122*

# W

wear particles  *56, 81, 95, 152-53, 183*

wheelchairs *29, 150*

# Z

Zirconia™ *68, 97, 154, 184*

Alpena Co. Library
211 N. First Ave.
Alpena, MI 49707

Made in the USA